Transdisciplinary Interfaces and Innovation in the Life Sciences

Medizin, Technik und Gesellschaft
Medicine, Technology and Society

Herausgegeben von Rainer Paslack, Frank Böker und Hilmar Stolte
Edited by Rainer Paslack, Frank Böker and Hilmar Stolte

Bd./Vol. 5

Zur Qualitätssicherung und Peer Review der vorliegenden Publikation

Die Qualität der in dieser Reihe erscheinenden Arbeiten wird vor der Publikation durch Herausgeber der Reihe geprüft.

Notes on the quality assurance and peer review of this publication

Prior to publication, the quality of the work published in this series is reviewed by editors of the series.

Eunsook Hyun / Rainer Paslack / Hilmar Stolte (eds.)

Transdisciplinary Interfaces and Innovation in the Life Sciences

PL ACADEMIC RESEARCH

Bibliographic Information published by the Deutsche Nationalbibliothek
The Deutsche Nationalbibliothek lists this publication in the Deutsche Nationalbibliografie; detailed bibliographic data is available in the internet at http://dnb.d-nb.de.

Library of Congress Cataloging-in-Publication Data
Transdisciplinary interfaces and innovation in the life sciences / Eunsook Hyun, Rainer Paslack, Hilmar Stolte (eds.).
p. ; cm. – (Medicine, technology, and society ; Bd./vol. 5)
ISBN 978-3-631-65266-4 – ISBN 978-3-653-04367-9 (e-book)
I. Hyun, Eunsook, 1962- editor. II. Paslack, Rainer, editor. III. Stolte, Hilmar, editor. IV. Series: Medicine, technology, and society ; v. 5. 0943-5131
[DNLM: 1. Biological Science Disciplines. 2. Biomedical Research. 3. Cooperative Behavior. 4. Diffusion of Innovation. 5. Interdisciplinary Communication. 6. Socioeconomic Factors. W1 ME655B v.5 2014 / W 20.5]
R850
610.72'4–dc23

2014013262

Printed with support of "IALS Life Sciences University Network", Potsdam-Berlin-Boston and "IFB-Tx Integrated Research and Treatment Center Transplantation", Hannover.

ISBN 978-3-631-65266-4 (Print)
E-ISBN 978-3-653-04367-9 (E-Book)
DOI 10.3726/978-3-653-04367-9

PL Academic Research is an Imprint of Peter Lang GmbH.

Peter Lang – Frankfurt am Main · Bern · Bruxelles · New York · Oxford · Warszawa · Wien

This publication has been peer reviewed.

www.peterlang.com

Preface

As an artifact of human invention, science is highly dynamic. Life science comprises any of the branches of natural science dealing with the structure and behavior of living organisms and their relationships to one another and the environment. In Transdisciplinary Interfaces and Innovation in the Life Sciences, an edited book, we present the manner in which the International Academy of Life Sciences' (http://www.lifesciences.net) biomedical mission, aim, commitment, international network, and cross-disciplinary engagement have expanded and deepened the dynamic field of life sciences. We value the perspective of Rudolf Virchow (1821–1902), who believed in "medicine as social science"; that is, medicine is and should always be a socially responsive human engagement advancing the condition of the living. From this perspective, the boundary between natural sciences and social sciences is and should be indistinct; thus the borderless transdisciplinary approach advances the field of life sciences.

The objective of IALS is to examine the interface between medicine and the life sciences. This necessarily incorporates ethics. IALS has examined this topic in detail in three transatlantic workshops. The first in Berlin (2010) concentrated on the interface between the expanding areas of molecular biology, nano-technology, synthetic biology, as well as the resulting ethical issues and the need for regulation. It became clear at this workshop that advancement of the new areas of life sciences needed a new approach shared between research universities, government and the commercial sector. This then became the subject for the second workshop (2011). The third, highly successful workshop (2012) continued the same theme, placing emphasis on the role of regulation and ethics in clinical research. The discussion was expanded to include risk management and the different approaches used in Europe and the USA. The regulation and ethics of tissue engineering ended with a discussion of the difficulties arising from conflict of interest and the effect on clinical research. Many of these issues have now been crystallized, and the role of IALS is to assess and debate how to progress in this complex interdisciplinary field.

This book is a collection of conference proceedings that represent years of collaboration between the International Academy of Life Sciences (IALS) in Germany and the University of Massachusetts Boston (UMass Boston) in the US.

We would like to acknowledge the many years of special intellectual support from Professor Robert Price, Ph.D., Kings' College, London, in the fascinating area of "Campus Life Sciences".

Potsdam, November 2013

Professor Eunsook Hyun, Ph.D.
Dr. Rainer Paslack, Ph.D.
Professor Hilmar Stolte, MD

Contents

Introduction

Taking a Transdisciplinary Approach to the Life Sciences

Eunsook Hyun

1. Introduction

A means to increasing integrated knowledge and practice without disciplinary boundaries, transdisciplinarity may ultimately provide an avenue for engagements promoting an insightful understanding and problem-solving capacity to complex human problems. Transdisciplinarity is a democratic epistemology and action (Hyun, 2011) that challenges all of us in academia, industry, government, and various human organizations, both public and private, the basis of whose operation comprises all forms of existing disciplinary human knowledge, science, and technology (e.g., Darbellay, Cockell, Billotte, & Waldvogel, 2008; Hadorn, Bradley, Pohl, Rist, & Wisemann, 2006; Kaufman, Moss, & Osborn, 2003; Kessel, & Rosenfield, 2008; Klein, 2004; McWilliam, Hearn, & Haseman, 2008; Nicolescu, 2002, 2007; Ramadier, 2004). The essential purposes of transdisciplinarity are (a) to signify and to engage in a unity of knowledge and surpass mono-, inter-, and multidisciplinary approaches (Hyun, 2011) and (b) to build a capacity for human knowledge that is responsive to solve problems affecting real lives.

Transdisciplinarity inspires us to ponder a unity of knowledge beyond traditionally accepted disciplines (e.g., Nicolescu, 2002). Jean Piaget, a French-speaking Swiss developmental psychologist known for his theory of cognitive development and epistemological studies (generic epistemology) with children, introduced the term transdisciplinarity in 1970. Throughout his life, Piaget argued that education is the only means to saving human societies from collapse, regardless of whether cataclysmic or gradual. A question, however, remains: What kinds of education might be introduced to learners, researchers, policy makers, and various practitioners that would lead to a unity of human knowledge capacity, practice, potentiality, and awareness beyond the disciplinary boundaries.

To uncover the nature and characteristics of the flow of information circulating among the various branches of knowledge, to develop research in a new scientific transdisciplinary mode, and to create a global competence yielding educational capacity (de-educate, re-educate, and newly educate) in borderless disciplines, the International Center for Transdisciplinary Research (CIRET) was founded in 1987 as a nonprofit organization located in Paris. The

Charter of Transdisciplinarity, adopted in 1994 at the 1st World Congress of Transdisciplinarity in Convento da Arrábida, Portugal, was the boldest and most significant move in transdisciplinarity, signifying social responsibility of both academia and nonacademic entities as well as their collective transformative actions on a global scale (see Table 1).

Table 1: The Charter of Transdisciplinarity

Preamble

Whereas, the present proliferation of academic and non-academic disciplines is leading to an exponential increase of knowledge which makes a global view of the human being impossible;

Whereas, only a form of intelligence capable of grasping the cosmic dimension of the present conflicts is able to confront the complexity of our world and the present challenge of the spiritual and material self-destruction of the human species;

Whereas, life on earth is seriously threatened by the triumph of a techno-science that obeys only the terrible logic of productivity for productivity's sake;

Whereas, the present rupture between increasingly quantitative knowledge and increasingly impoverished inner identity is leading to the rise of a new brand of obscurantism with incalculable social and personal consequences;

Whereas, an historically unprecedented growth of knowledge is increasing the inequality between those who have and those who do not, thus engendering increasing inequality within and between the different nations of our planet;

Whereas, at the same time, hope is the counterpart of all the afore-mentioned challenges, a hope that this extraordinary development of knowledge could eventually lead to an evolution not unlike the development of primates into human beings;

Therefore, in consideration of all the above, the participants of the First World Congress of Transdisciplinarity (Convento da Arrábida, Portugal, November 2–7, 1994) have adopted the present Charter, which comprises the fundamental principles of the community of transdisciplinary researchers, and constitutes a personal moral commitment, without any legal or institutional constraint, on the part of everyone who signs this *Charter*.

Article 1:

Any attempt to reduce the human being by formally defining what a human being is and subjecting the human being to reductive analyses within a framework of formal structures, no matter what they are, is incompatible with the transdisciplinary vision.

Article 2:

The recognition of the existence of different levels of reality governed by different types of logic is inherent in the transdisciplinary attitude. Any attempt to reduce reality to a single level governed by a single form of logic does not lie within the scope of transdisciplinarity.

Article 3:
Transdisciplinarity complements disciplinary approaches. It occasions the emergence of new data and new interactions from out of the encounter between disciplines. It offers us a new vision of nature and reality. Transdisciplinarity does not strive for mastery of several disciplines but aims to open all disciplines to that which they share and to that which lies beyond them.

Article 4:
The keystone of transdisciplinarity is the semantic and practical unification of the meanings that *traverse* and *lay beyond* different disciplines. It presupposes an open-minded rationality by re-examining the concepts of "definition" and "objectivity". An excess of formalism, rigidity of definitions and a claim to total objectivity, entailing the exclusion of the subject, can only have a life-negating effect.

Article 5:
The transdisciplinary vision is resolutely open insofar as it goes beyond the field of the exact sciences and demands their dialogue and their reconciliation with the humanities and the social sciences, as well as with art, literature, poetry and spiritual experience.

Article 6:
In comparison with interdisciplinarity and multidisciplinarity, transdisciplinarity is multireferential and multidimensional. While taking account of the various approaches to time and history, transdisciplinarity does not exclude a transhistorical horizon.

Article7 :
Transdisciplinarity constitutes neither a new religion, nor a new philosophy, nor a new metaphysics, nor a science of sciences.

Article 8:
The dignity of the human being is of both planetary and cosmic dimensions. The appearance of human beings on Earth is one of the stages in the history of the Universe. The recognition of the Earth as our home is one of the imperatives of transdisciplinarity. Every human being is entitled to a nationality, but as an inhabitant of the Earth is also a transnational being. The acknowledgement by international law of this twofold belonging, to a nation and to the Earth, is one of the goals of transdisciplinary research.

Article 9:
Transdisciplinarity leads to an open attitude towards myths and religions, and also towards those who respect them in a transdisciplinary spirit.

Article 10:
No single culture is privileged over any other culture. The transdisciplinary approach is inherently transcultural.

Article 11:
Authentic education cannot value abstraction over other forms of knowledge. It must teach contextual, concrete and global approaches. Transdisciplinary education revalues the role of intuition, imagination, sensibility and the body in the transmission of knowledge.

Article 12:
The development of a transdisciplinary economy is based on the postulate that the economy must serve the human being and not the reverse.

Article 13:
The transdisciplinary ethic rejects any attitude that refuses dialogue and discussion, regardless of whether the origin of this attitude is ideological, scientistic, religious, economic, political or philosophical. Shared knowledge should lead to a shared understanding based on an absolute *respect* for the collective and individual Otherness united by our common life on one and the same Earth.

Article 14:
Rigor, *openness*, and *tolerance* are the fundamental characteristics of the transdisciplinary attitude and vision. *Rigor* in argument, taking into account all existing data, is the best defense against possible distortions. *Openness* involves an acceptance of the unknown, the unexpected and the unforeseeable. *Tolerance* implies acknowledging the right to ideas and truths opposed to our own.

Article final:
The present *Charter of Transdisciplinarity* was adopted by the participants of the first World Congress of Transdisciplinarity, with no claim to any authority other than that of their own work and activity.
In accordance with procedures to be agreed upon by transdisciplinary-minded persons of all countries, this *Charter* is open to the signature of anyone who is interested in promoting progressive national, international and transnational measures to ensure the application of these Articles in everyday life.

Convento da Arrábida, 6th November 1994
Editorial Committee
Lima de Freitas, Edgar Morin and Basarab Nicolescu
Translated from French by Karen-Claire Voss

Source: CIRET (1997)

2. Transdisciplinarity and the *Triple T*

Transdisciplinarity does not entail new disciplinary knowledge; instead it involves an inherent and perpetual capacity of simultaneously deepening and extending disciplinary approaches for the improvement of borderless human knowledge and engagement. It inspires or triggers the emergence of new or different information and interactions derived from encounters among various disciplines and leads to new and different possibilities of understanding nature and engaging with reality. Transdisciplinarity does not require mastery of disciplines, but it motivates individuals to work toward openness by *dialoguing, sharing, connecting, fusing,* and *surpassing* existing disciplinary boundaries. In this borderless process, "*no single culture is privileged over any other culture. The transdisciplinary approach is inherently transcultural*" (as indicated in the Article 10). Basarab Nicolescu (1994) claims only transdisciplinary approach can effectively reconcile the twenty-first century's challenges. He further argues, "*universal sharing of knowledge cannot take place without the emergence of a new tolerance founded on the transdisciplinary attitude,*" suggesting the necessity of putting transcultural, trans-religious, trans-political, and transnational visions into the transdisciplinary practice (p. 1).

Recalling Piaget's impulse toward transformative education and with its underlying responsibility for humanity, I believe contemporary human knowledge formation, education, and engagement must be based on three pillars: the transdisciplinary, the transcultural and the transnational (Hyun, 2011). I call these the *Triple T*, through which ethical practitioners intrinsically reject any attitudes that preclude open dialogue and interaction, regardless of whether the origin of such an attitude is ideological, "scientistic", religious, economic, geopolitical, philosophical, or even oriented toward a social-emotionally internalized discipline. Integrated open dialogue, discussion, and interaction without an artificial border should lead to the formation of shared new knowledge with mutually understandable language, which will reduce and ultimately eliminate an uncomfortable, unfortunate, and unfair condition of "otherness," and increase integrated collective knowledge to use in all forms of problem solving in human communities (Nicolescu 2007, 2008). To that end, *respect* for one another, *rigor* in disciplinary knowledge coupled with interconnected and interdependent interaction for unity of knowledge, *openness* to self and disciplinary doubts, and *acceptance* of various ways of knowing are the essential elements of the Triple T guided by transdisciplinarity and its form of engagement. Transdisciplinarity enhances borderless and cross-border academic leadership, education, research, and service to all forms of local, national, and international constituencies and advances humanity for social and public good, leading to global significance.

The transdisciplinary approach is usually taken with the explicit intent to solve complex and multidimensional social-ecological problems, such as those involving the interface of humans, their health, society, and its natural systems. Thus, life scientists in particular are poised to seriously consider adapting and infusing transdisciplinarity into their research and development (R&D), teaching, learning, and public service.

3. Zooming in on Life Sciences and the Need for the Transdisciplinary Approach

Life sciences involve the study of the diversity and dynamics of living systems from biology to molecular biology to neurobiology to stem cell research to biotechnology to biomedical engineering; from chemistry to biochemistry to evolutionary biology; and from bio-ethics to public policies (e.g., public health policy, environmental health policy) and global governance for human rights. Thus, the responsive field of the life sciences represents the evolving interface of humans, their society, and its biological and natural systems in both contemporary and futuristic modes.

Twentieth-century Western monodisciplinary fields (e.g., biology, chemistry, physics, philosophy, religion, and psychology, based on Aristotelian and Cartesian philosophy, which included an isomorphic, reductionist perspective on the world) have actively become methodologically linked with other fields and known as interdisciplinary fields, and in some cases have generated new disciplinary knowledge, such as biochemistry, neurobiology, biomedical engineering (Magner, 2002). Another example, integrative biology is the study and research of biological systems. It does not simply involve one discipline but integrates a wide variety of disciplines that methodologically work together to find answers to scientific questions: Biological sciences, engineering, physical sciences, and social sciences are all incorporated to address specific biological life science issues. In the 21st century, however, we are no longer interested in simply exploring super-monodisciplinary scientific inquiries in a methodologically integrated fashion to advance the knowledge (e.g., stem cell engineering) but face complex real-life application, impact, and problems solving (equity, disparity, accessibility, protection) in advancing life science engagements *in, for*, and *with* public intersecting with ethical principles, dilemmas, challenges, and difficulties. Management, public policy, and global governance for both the protection and enhancement of basic human rights *are* and *must be* inherent and omnipresent characteristics in life sciences. Life sciences-oriented problem solving operates in many cases in conjunction with health sciences (e.g., epidemiology) and social sciences (e.g., sociology, psychology, education, management, public policy, political science, global governance, philosophy, and human rights and securities). The 21st century's new players in life science research innovation and development have begun utilizing transdisciplinary approach for futuristic endeavor (e.g. Mroczkowski, 2012: 277).

Since World War II, the global population, crop production, and energy consumption have increased dramatically. In addition, the total amount of human knowledge has accumulated in equally dramatic fashion – mass production, mass consumption, mass disposal, mass mobility, mass environmental impact, mega-urbanization, complex public health problems, and tremendous educational disparity between poor and rich. Scientists today often do not understand what is going on in their neighbor's laboratory. They cannot even communicate one another: They may pursue the same topics of research but in different languages that are specific to only their own disciplinary thinking. The phenomenon of compartmentalization poses the greatest obstacle to the initiation of transdisciplinary knowing. When we confront a real-life issue that demands resolution, we need to educate ourselves to combine or integrate disciplines to understand and solve problem along with educating students in this mode for the societal sustainability. If the problem were one that could be handled by a signal discipline, it would have been solved already or it would be solved more easily sooner than later (Sumi, 2008). How this objective is attained in practice? Who is supposed to contribute to process of attaining it? In what ways can universities facilitate this process? How can academia provide support and lead in this transformative engagement in collaboration with industry, governmental and non-governmental organizations, and various public and private entities?

4. The Evolution of Collaboration between International Academy of Life Sciences (IALS) and University of Massachusetts Boston

With the fusion of health sciences and social sciences, since 2008 UMass Boston and IALS have worked together to advance transdisciplinary engagement in life sciences among academia, industries, governmental and nongovernmental sectors, students, and various public and private partners. As of 2012, we have had three annual collaborative workshops (2010, 2011, and 2012); topics derived from the theme of *Advancing Public Policy: Innovation and Risk in Life Sciences*.

(1) In 2010, the 1st annual IALS-UMass Boston workshop announcement stated: Workshop "Transdisciplinary Interfaces of Molecular Biology, Nanotechnology, and Bioengineering: International Regulations, Ethics and Public Issues." February 12th–13th, 2010, Langenbeck-Virchow-Haus, Berlin, Aesculap Academy.

I. Questions to be answered
Transdisciplinary research is usually aimed explicitly at complex social and ecological problems, such as those at the interface of humanity, society and natural systems. Transdisciplinary approach implies full interaction between disciplines from a problem-based perspective. Transdisciplinary vision is transnational and transcultural, encompassing a wealth of ethics, beliefs and creative impulses.

II. Needs
Scientific institutions, medical organizations and companies involved in the Life Sciences, especially in the area of pharmaceutical and clinical research, operate within a legal environment which is controlled by international and national laws

and ethical standards. These bodies all have to be able to identify the key statutes and trends in international law as they relate to the Life Sciences. Furthermore, gaining social and public acceptance essentially depends on attention to different ethical values within clinical research (e.g., stem cell or genome research). This is a prerequisite for the successful marketing of life science-based products, such as pharmaceuticals.

III. Background

Biotechnology, nanotechnology, bioinformatics and synthetic biology within the life sciences, are all areas of special interdisciplinary interest. Biotechnology is often referred to as genetic engineering and combines disciplines like molecular biology, genetics, cell biology and bioinformatics. Controversial questions have been raised, e.g. the ownership and control of genetic information. At an individual level, the absence of legal protection can lead to discrimination in employment or insurance.

Recently there has also been much debate about the future implications of nanotechnology, which has the potential to create many new materials and devices, for example in medicine. They range from nanomaterials to nanoelectronic biosensors to possible applications for molecular nanotechnology.

Synthetic Biology refers to the design and fabrication of biological components that do not already exist in the natural world.

(2) In 2011, the 2nd annual IALS-UMass Boston workshop announcement stated:

Workshop "Advancing in the Life Sciences and Uniting Universities, Corporate and Government Partners." October 27–29, 2011, Healey Library, University of Massachusetts Boston, Boston, Massachusetts, USA.

We are excited to announce the date and location of the next in our ongoing series of Transdisciplinary Life Sciences conferences and would like you to mark your calendars for October 27 through 29, 2011. This conference will take place in Boston, Massachusetts as a direct follow up to the international workshop hosted by International Academy of Life Science (IALS) and the Life Sciences University Network in Berlin, Germany[1] in February 2010.

The 2010 conference gathered international experts from various state-of-the-art life sciences areas to give talks, which would inspire transdisciplinary as well as transcultural collaborations, and focused on the theme 'Transdisciplinary Interfaces

---

1 http://wiki.acadlife.net/wiki/Main_Page.

of Molecular Biology, Nanotechnology, and Bioengineering: International Regulations, Ethics and Public Issues.' The conference brought together diverse transdisciplinary national and international approaches to topics such as nanotechnology, bioinformatics and synthetic biology, ethics, and medical sociology.

2011 Conference Purpose

Building on the outcomes of the first IALS conference, the 2011 conference will bring internationally renowned experts from academia, government, non-profit, and the private sector to discuss the advancement of common goals, objectives, and global interaction in the life sciences. The participants will identify and explore opportunities for the creation of transnational, transdisciplinary and cross-sectoral partnerships linking higher education institutions, corporations and public agencies in life sciences across the globe.

Experts with wide-ranging perspectives and experience will review issues facing the global biosciences community, including:

- Increasingly significant role of regional university research in driving innovation in the global biosciences arenas; and
- how universities could balance their role as an important regional research and cultural actor, with the increasing demands for global sharing and transfer of knowledge and technology.

The panel discussions will also expand on the concept of a "biosciences benefit circle" that describes a nonlinear movement of information and research sharing among four key stakeholder groups: universities, market-facing entrepreneurs and businesses, government and regulatory agencies, and NGOs and advocacy organizations. The discussion will review the role of universities as key integrators driving interactions and cooperation among multiple stakeholders, across disciplines and cultures.

(3) In 2012, the 3rd annual IALS-UMass Boston workshop announcement stated:

Workshops entitled "Regulation and Ethics in Clinical Research" and *Advancing Public Policy: Innovation and Risk in Life Sciences – Bridging Public Policy and Innovation in the Life Sciences* August 22–25, 2012 *Schloss Cecilienhof, Potsdam Brandenburg, Germany*

This year's conference combines the first of a new seminar series with our ongoing transnational and transdisciplinary dialogs on the important issues facing Life Sciences professionals worldwide.

In cooperation with the Medizinische Hochschule Hannover, the IFB-Centres of Germany, and PRIM&R in Boston, we are launching a new seminar series specifically designed for clinical scientists.

This first seminar "Regulation and Ethics in Clinical Research" will enhance clinical scientists' theoretical and practical skills for clinical investigations. The seminar will offer three training sessions, from Wednesday, 22 August through Thursday, 23 August.

Seminar Objectives: The Clinical Scientist Training Program is committed to educating and training highly motivated individuals to become successful, independent clinical investigators and future leaders in academic medicine and biomedical research. It is designed primarily for junior faculty and senior subspecialty fellows. This seminar is designed to enhance clinical scientists' theoretical and practical skills for clinical investigations. It is designed to be followed up by the Workshop "Advancing Public Policy: Innovation and Risk in Life Sciences." The focus will be on Exploring and Examining Ethical Issues and Regulatory Requirements Central to Clinical Research. Our partnership with PRIM&R reflects their roles in educating and engaging the public and the research community, and as a global resource on research ethics. The seminar offers three training units (½ day each).

Workshop *Advancing Public Policy: Innovation and Risk in Life Sciences—Bridging Public Policy* and Innovation in the Life Sciences. This workshop is an ongoing initiative of the International Academy of Life Sciences (IALS) in cooperation with the University of Massachusetts, Boston. IALS has invited numerous experts from state-of-the-art disciplines to inspire transdisciplinary as well as transcultural, discussion (see http://wiki.acadlife.net/wiki/Main_Page). This workshop will be a full day, Friday, 24 August, followed by a day of cultural activities in the Potsdam area. We would like to invite you to join us in our ongoing series of IALS transatlantic, transdisciplinary conferences, which began with Technology Transfer and Bioethics back in the 1980s. Under the umbrella of the "IALS Life Sciences University Network" IALS continues to reflect the challenges of our discipline, and has expanded to include a role as a global network of Life Sciences experts.

Scientific institutions, medical organizations and companies involved in the life sciences, especially in the area of pharmaceutical and clinical research, operate within a legal environment which is controlled by international and national laws and ethical standards. These bodies all have to be able to identify the key statutes and trends in international law as they relate to the life sciences. Furthermore, gaining social and public acceptance essentially depends on attention to different ethical values within clinical research (e.g. stem cell or genome research). This is a prerequisite for the successful marketing of life-science-based products, such as pharmaceuticals.

*

As the three years of annual workshop conference program announcements indicate, our transdisciplinary, transcultural, and transnational engagement have evolved and transformed our collective thinking (see chapters written by Radden and Hannigan as example). A book entitled *Transdisciplinary Interfaces and Innovation in the Life Sciences* represents three years' dialogue among the participants. We hope that this work inspires further sharing of responsibility and borderless engagement in life sciences, health sciences, social sciences, and beyond for the sake of humanity.

5. Conclusion

In comparison with other disciplinarities (i.e., monodisciplinarity, interdisciplinarity, multidisciplinarity, etc. See elsewhere, Hyun, 2011; Klein, 2004; Tress, Tress, & Fry, 2006), transdisciplinarity involves individuals from different academic disciplines working together without borders – practitioners and nonacademic laypersons from areas including industry, business, public administration, nongovernmental organizations – to understand real-life issues and phenomena and to collectively solve the problems. Some universities have a transdisciplinary academic department or school as an academic unit with degree programs. It requires more than academic departments' borderless engagement, more than creating new academic units (departments, schools, colleges, universities). The transdisciplinary approach is a relatively new form of *shared* research, *shared* learning, *shared* teaching, *shared* interaction, and *shared* problem solving for ultimately *shared* social responsibilities involving different parts of the public and academia in order to meet the complex challenges of human society and its sustainability.

At the macro level, the transdisciplinary approach seeks to promote the meaningful collaboration of academia with society's various constituents (industry, governmental organizations, nongovernmental organizations, the citizenry) in order to produce unified knowledge to understand and to solve complex problems of local, regional, and global scope. Improving basic human rights, preserving ecology and its systems, and dealing with the pressing issues of cultural, economic, and environmental survival demand rigorous, relevant, unbiased, and collectively engaged forms of research and action. All educational opportunities must include results and insights of various forms of scholarship, knowledge and problem-solving that emerge at the intersection of disciplines, academia, and society. We need to critically rethink the production of academic knowledge and education, starting with (a) identifying how they contribute to understanding and solving complex societal problems and (b) learning how to coproduce knowledge with society and how to navigate and communicate between the different modes of knowledge production to optimize these contributions. As the work of both Jean Piaget (cognitive psychologist) and Basarab Nicolescu (theoretical physicist) constantly remind us, these challenges require a dedicated investment in transdisciplinary teaching, learning, research and outreach – the collective education process for all.

References

CIRET: International Center for Transdisciplinary Research (1997): Retrieved from http://basarab.nicolescu.perso.sfr.fr/ciret/index_en.php

Darbellay, F., Cockell, M., Billotte, J., & Waldvogel, F. (eds.) (2008): *A vision of transdisciplinarity: Laying foundations for a world knowledge dialogue.* Boca Raton, FL: Taylor and Francis Group.

Hadorn, G. H., Bradley, D., Pohl, C., Rist, S., & Wisemann, R. (2006): Implications of transdisciplinarity for sustainability research. *Ecological Economics, 60*, pp. 119–128.

Hyun, E. (2011). Transdisciplinary higher education curriculum transformation: A complicated cultural artifact. *Research in Higher Education Journal,11,* pp. 1–19. Retrieved from http://www.aabri.com/manuscripts/11753.pdf

Kaufman, D., Moss, D., & Osborn, T. (eds.) (2003): *Beyond the boundaries: A transdisciplinary approach to learning and teaching.* Praeger: Westport, CT.

Kessel, F., & Rosenfield, P. (2008). Toward transdisciplinary research: Historical and contemporary perspectives. *American Journal of Preventive Medicine, 35(2S)*, pp. S225–S234.

Klein, J. T. (2004): Prospects for transdisciplinarity. *Futures, 26*, pp. 515–526.

Magner, L. (2002): *A history of the life science*, 3rd ed. New York: CRC Press.

McWilliam, E., Hearn, G., & Haseman, B. (2008): Transdisciplinarity for creative futures: What barriers and opportunities? *Innovations in Education and Teaching International, 45*(3), pp. 247–253.

Mroczkowski, T (2012): *The new players in life science innovation: Best practices in R&D from round the world.* Upper Saddle River, NJ: FT Press.

Nicolescu, B. (1994): *The transdisciplinary evolution of learning.* International Center for Transdisciplinary Research (CIRET).

Retrieved from http://www.learndev.org/dl/nicolescu_f.pdf

Nicolescu. B. (2002): *Manifesto of transdisciplinarity* (K.-C. Voss, Trans.). New York, NY: SUNY Press.

Nicolescu, B. (2007): *Transdisciplinarity and the unity of knowledge: Beyond the science and religion dialogue.*

Retrieved from http://www.metanexus.net/magazine/tabid/68/id/10013/Default.aspx.
Nicolescu, B. (ed.) (2008): *Transdisciplinarity: Theory and practice.* Cresskill, NJ: Hampton Press.
Ramadier, T. (2004): Transdisciplinarity and its challenges: The case of urban studies. *Futures, 36*, pp. 423–439.
Sumi, A. (2008): New initiatives of the University of Tokyo toward establishing a strategy for sustainability through knowledge structuring and a transdisciplinary approach. In: F. Darbellay, M. Cockell, J., Billotte, & F. Waldvogel (eds.): *A vision of transdisciplinarity: Laying foundations for a world knowledge dialogue* (pp. 167–172). Boca Raton, FL: Taylor and Francis Group.
Tress, B., Tress, G. & Fry, G. (2006): Defining concepts and the process of knowledge production in integrative research. In: B. Tress, G. Tress, G. Fry, & P. Opdam (eds.), *From landscape research to landscape planning* (pp. 13–26). Dordrecht, Netherlands: Springer. Retrieved from http://library.wur.nl/frontis/landscape_research/02_tress.pdf.

Transdisciplinarity and Ethics as Challenges for Life Sciences

Transdisciplinarity from a Problem-Based Perspective

Winston Langley

Good morning, to you![1]

Allow me to begin by saying I did not select the subject which I am about to begin discussing with you. I say this, not as an accusation, but in the spirit of indicating that I would not, in general, have chosen for myself a subject of such intellectually demanding reach; I would rather have allowed it to more distinguished colleagues gathered here. It was assigned to me, however, and as an assignee, I accepted – as assignees are apt to do.

I hope you will, therefore, help me in the presentation, through the insights you can offer, as you make your comments and pose your questions. The approach the presentation will observe is as follows: it will seek to define how the term transdisciplinarity is being used; to look at certain existential conditions that have brought it to recent prominence; to touch on the history of our current, dominant disciplinary orientation in research and teaching; to look at a constitutive feature of the life sciences; and then to point to a central attribute of the nature of knowledge; and, finally, to offer a conclusion.

1. Definition

The second half of the twentieth century, especially its last two-fifths, saw a focus on what has come to be called multi-, pluri- and interdisciplinarity, with the former two emphasizing the implications of several disciplinary foci on a subject area, while the latter focus underlined the relationship between and among disciplines. A central feature of those foci is that they have not really sufficiently called into question disciplinary boundaries or practices; on the contrary, they have sought to preserve those boundaries.

Transdisciplinarity, however, is an approach to learning and research that transgresses disciplinary borders, seeking as it does to effect articulation through and across disciplines and to pursue joint or common beliefs, practices, reasons, axioms,

1 Note: this speech is made at the first IALS-UMass Boston Conference in Berlin, Germany, 2010.

systems, "goings-on," and languages than can capture the complex relationships that exist between and among phenomena.[2]

Understood as just defined, transdisciplinarity does not have as its practice or goal the preservation of disciplinary boundaries as we have come to know, use, depend on and – perhaps – love them. Neither does it seek to protect – and disciplines do protect – the cultural or civilizational outcomes that these disciplines produce or on which they ground themselves. As such, it is both an approach to research and learning and a certain form of action-reaction-orientation.

2. Existential Conditions

The emerging focus on transdisciplinarity is not the product of either an immaculate or parthenogenic conception. It has evolved out of certain existential conditions, which human beings have been facing, on an increasing scale. These conditions have included the development of what we have come to call the knowledge-based society – one consciously promoted and experienced, and requiring high levels of predictability and routine; the emphasis on technology, as a form of planned socio-economic and technical development, with its ideological undertones of modernity and progress; the maturation of almost an idolization of intellectual exactitude which, among other things, espoused the "finality of scientific conclusions"; the collaboration of specialists – reflecting disciplinary and sub-disciplinary thinking that sponsored deep but narrow levels of expertise and the resulting compartmentalization of knowledge, as well as its attendant psychology of "security" within those compartments; and the evolution of a secular order that denied human spirituality or made spirituality co-extensive with religion.

There has been another side to those just-recited conditions, however. "Unruly complexity"[3], in the form of the environment, global warming, poverty, non-linear dynamics, hybridity, and uncertainties began to "confuse" the certainties modernity bequeathed, to identify where existing disciplines could not even find answers to the most pressing problems, and to point out where overlapping

---

2 See, among others, Julie Thompson Klein, "Prospects for Transdisciplinarity" in: *Futures,* vol. 36 (2004), pp. 515–526; Thierry Ramadier, Transdisciplinarity and It's Challenges: the Case of Urban studies" in: *Futures* vol. 36 (2004), pp. 423–436; Manfred A. Max-Neef, "Foundations of Transdisciplinarity" in: *Ecological Economics*, vol. 53 (2005), pp. 5–16; John Berger, *Ways of Seeing* (New York: Penguin books, 1972); Niels Bohr, *The Unity of Knowledge* (New York: Doubleday and Co., 1955).

3 Peter Taylor, *Unruly Complexity: Ecology, Interpretation, Engagement* (Chicago: University of Chicago Press, 2005).

levels of reality subverted sacred conceptual frameworks. Indeed, to compound the problems we face, the narrow disciplinary frameworks allowed for no language that could sustain dialogue between and among those levels of reality beyond boundaries. Your own Werner Heisenberg, that deservedly decorated leader in quantum physics, taught us that while science "rests on experiments," it is "rooted in conversations."[4] Those conversations are what will show the relationships between the movement of electrons and "The Dance," which Matisse offered us, between morphemes and genes, and identify ways to improve our ecosystems and human welfare. In short, we will not only be able to name the broad interrelationships between biophysical reality and what might be generally called the human dimension, but, as well, to integrate the spatial and temporal, human, policy, and technical dimensions, and do that integration at the levels of plot, households, and watersheds or community," with community understood to connote our widest conception of it.

Transdisciplinarity seeks to give us the capacity for sustained dialogue across existing disciplinary borders, to help us understand that, while nature and society cannot be properly or neatly fitted into what passes for our exact and rational, disciplinary categories, we can nonetheless live relatively free of anxiety, despite uncertainty, ambiguity, and multilevel reality.[5]

3. History and Current Dominant Orientation

Although guilds, religious orders, scientific societies, research institutions, and other entities have taken turns in shaping modern science (and its offspring technology), universities, and, of late, research universities, have been the leading institutions constituting our modern body of knowledge. And the cultural West, which has dominated the modern period, has exported much of its approach to research and learning to most of the rest of the world. The latter, with modifications has continued the tradition.

Much of the modernist project – which launched the single disciplinary boundaries we are discussing – began in the seventeenth century, when the scientific order inaugurated by Cartesian logic out-distanced its rivals. That logic sought certainty; and in the latter's pursuit, advanced a defined rational method, insisted

4 Louisa Gilder, *The Age of Entanglement: when Quantum Physics was Reborn* (New York: Alfred A. Knolf, 2008), p. xiii.

5 Stephen Toulmin, *Cosmopolis: the Hidden Agenda of Modernity* (Chicago: University of Chicago Press, 1990).

on the requirement of exact language, and espoused the goal of unified science. All three became part of a single project.

Universities, claiming to embody or to be pursuing excellence, advocated the virtue of knowledge which was *stable* and *predictable*; and knowledge which did not partake of those characteristics could not be regarded as scientific. (Here we must bear in mind that universities are not simply sites of new learning or dissemination of existing knowledge, but also the principal conservators of knowledge, with professors and students serving as the leading agents of that conservation).[6]

The stable, the predictable, as well as the controllable – as earlier said – also provided the environment for narrower and narrower components of knowledge, often bearing little or no relationship to social life. The relative cultural homogeneity of the nation-state, which also evolved in the seventeenth century, reinforced the notion of the stable and the predictable, even as the evidence of increased cultural heterogeneity of international and transnational interactions made stability uncertain and unpredictable. The shifts to inter- and pluridisciplinary studies have been attempts to deal with the challenges of complexity and heterogeneity. As indicated above, however, those shifts have not even sought to question disciplinary borders.

The idea of the unity of science (and some would say the unity of knowledge), is an admirable, admirable goal, but it temps the bringing into being of a less open system of knowledge and the marginalization of that which does not now conform to existing scientific knowledge. One may have it as an ultimate goal (and it may even be part of what has been called the "implicate order," which one should constantly bear in mind). For the time being, however, I share the view of some thinkers that the goal of transdisciplinarity should be coherence – not unity.

It is under the auspices of the emphasis on the rational (later to become the rational and the experimental) method, exact language, and unity of sciences that the disciplines – as we know them today – evolved. The life sciences, for example, covers disciplines and research/teaching programs extending from biology, biochemistry, chemistry, healthcare and healthcare administration, as well as population genetics and nanotechnology, to psychology, environmental biology, bio-informatics, physics, and environmental studies, among others.

The final feature of this history is the philosophy that has accompanied the approaches to scientific knowledge – positivism, post-positivism, and constructivism. In the case of the first approach (which, as you know, helped define and

6 Michael Oakeshott, *Rationalism in Politics as Other Essays* (Indianapolis: Liberty Press, 1991), pp. 184–218; see, also, Toulmin, op.cit., pp. 104–105; 181–184.

justify much of the disciplinary compartmentalization with which we have been dealing), it claims that there is a reality external to human beings and human strivings, but human beings can come to recognize this reality through experimental research. Guided by this experimentation, human beings do not only come to recognize but identify and elaborate universal laws, untethered to time, place, or cultural formations.

Post-positivism, which also acknowledge the idea of reality external to humans, takes the position that this reality is sufficiently complex that, in addition to the experimentation, one needs a sort of inter-subjective agreement among scientists, because human senses are even more unreliable than even the positivists suppose. The two schools of philosophy, especially the first, are associated more with what we call the exact sciences. There is another school – that called constructivism.

Embraced by most humanists and social scientists, this school argues that social reality is constructed, which is to say, that there is no social reality independent of the observer; thus, social scientists and humanists should be concerned about understanding the meaning people ascribe to reality. In short, there is no social reality outside that which human beings help to shape. As such, elaborating universal laws, independent of the meanings people attribute to them, might undermine understanding, because constructed reality is not independent of cultural formation, either. Do we hear quantum physics?

4. The Life Sciences

The life sciences, which have brought us here, present us with a special challenge, as well as opportunity. The challenge resides in the fact that, as human beings, as living beings, as part of the animal kingdom, we are subject to nature's laws. We, however, also live in a socio-cultural world – one in which we interact with and respond to symbols and all the meanings attributed to those symbols. So, human beings and their behavior are subject to and expressive of the natural environment – its bio- and physio-chemical laws, as well as the cultural symbols which have come to define what we have called civilization. Saying that human beings are subject to or expressive of the natural environment, however true, runs the risk of being misunderstood. We can say something more directly: human beings are part of nature, and social and other life depends on the uninterrupted functioning of natural systems. Likewise, civilization itself has had its shape and direction (in the form of artistic and scientific achievements) from nature. The life sciences, therefore, offer us an opportunity, to use the challenge it poses to demonstrate the liberating possibilities of transdisciplinarity.

One way of embracing that opportunity is to find means by which biology (indeed the life sciences, which represent complex structures or systems) can share with the humanities and the social sciences some converged or joint modes of analyses to gain access to higher levels of complexity and more comprehensive ranges of understanding. (Perhaps even moving to overcome the unhelpful and outmoded divide between nature and culture, not to mention the *Two Cultures* which C. P. Snow wrote so eloquently about during the last century.) For instance, Roger D. Masters, in his work, *The Nature of Politics,* urged us to recognize that "any general theory of human life as a biological phenomenon seem to rest on the properties of complex systems that are capable of encoding, transmitting, and processing information".[7] We could, as he did, extend this further.

Because each phenotype must fall within a reaction range conditioned by its genotype (interacting with the environment) the gene pool of a species could fruitfully be conceptualized as a system of information storage, processing, and transmission. Cultural roles or norms establish, in a roughly analogous fashion, the reaction range within which individual personalities fall, depending on certain specific experiences to which each human being is exposed. Culture or tradition (or the inventory of cultural symbols), like genetic inheritance, is a means of storing, processing, and transmitting information concerning adaptive responses to varied contingencies. And the teaching-learning-memory of individuals and groups create the likely range of reaction to future environmental conditions and stimuli, both natural and cultural.[8]

If the preceding were to be tentatively accepted, we could then say that the informational systems of the human gene pool, a human culture (symbol pool) and a human individual or group do not exist in a vacuum; each is, rather, an analytically separable component of a concrete, complex physical system: the species *Homo sapiens* on the planet Earth, a human population adapted to a particular environment and ecological niche, and a human phenotype with its unique history."[9]

Even more important, these three levels of information, each corresponding to a different level of physical system, form a nested series: each human population is part of the species, just as each individual is part of the population. Similarly, the three information-coding systems form a nested set: cultural norms must lie within the potentialities of the human gene pool, just as individual learning must lie within the potentialities of the individual's culture and genotype. Yet neither

7 Roger D. Masters, *The Nature of Politics* (New Haven: Yale University Press, 1989), p. 132.
8 Ibid.
9 Ibid., p. 132–133.

the cultural norms nor individual learning can simply be reduced to the information coded in the human chromosomes, for both verbal language and the central nervous system can function within limits as autonomous, open information systems.[10]

So far, our focus has either avoided or touched on – by implications, only – the "inner world" that defines human beings: the spiritual world. This is one which thirsts for knowledge and meaning, which psychologically responds to mystery and invitations to intuitive leaps, and which invites our search for a measure of control and liberation in living. Goethe, who was not fully at home with the demonstrable limitations of Newtonian physics, sought to teach us that "science is *as much* [emphasis mine] an inner path of spiritual development as it is a discipline aimed at accumulating knowledge of the physical world. It involves not only a rigorous training of our faculties which can attune us to the spiritual dimension that underlies and interpenetrates the physical: faculties such as feeling, imagination, and intuition."[11]

Unless this inner world is explored through observation grounded on the most exacting training and linked to the evolution of normative structures and the symbol pool involved, the cross-fertilization between positivism/neo-positivism, on the one hand, and constructivism on the other (what may be called a form of quantum logic) will not be possible. There is more to the problems which transdisciplinarity must address.

At the root of human spiritual development to which Goethe referred is the evolution and nature of consciousness – one beyond the information storage and non-conscious processing, although this type of processing can lead us to certain helpful patterns. Matthew Arnold, in his poem "Dover Beach," was getting at this consciousness when he penned his "Let us be true to one another," because the world seemed so various, beautiful and new, had "really neither joy, nor love, nor light/nor certitude, nor peace, nor help from pain [...]" It is the horror of cruelty, the rapture of discovery and understanding, and the deep sadness of parting, profoundly felt or otherwise experienced from great literature (including medical narratives) across cultures and time that tells us that the humanities, the social sciences, and the life sciences can find bridges to the universal and to each other.

10 Ibid., p. 133.
11 Max-Reef, op. cit., p. 10.

5. Nature of Knowledge

Much earlier in this presentation, I referred to existential conditions which have invited the emerging focus on transdisciplinarity. There is a fundamental feature of the human constitution that must be associated with that focus – something the late Michel Foucault informed us about: that "human fulfillment requires an opening up of possibilities" that reside beyond the limits of prevailing norms, beyond regularity, existing practices, comforts, consoling givens, beyond the sweetness of life and living.[12] This opening, which is part of the spiritual search for knowledge (as well as freedom), additionally, in moral terms, requires transgressive ruptures in the use of knowledge as an agent in the transformation of human life and the human condition. So, public policy will need us to enquire into that part of the person which constitutes the true domain of ethical judgment and the ways a person can best organize his or her relationship to moral obligation.

Transdisciplinarity is part of the human struggle for fulfillment, "to inhabit an ever more intelligible or less mysterious world"; and, as such, no fact has any finality or authority "over further adventures in understanding"; they are but contingent starting-places of and for further critical inquiry.[13]

This has been part of the history and nature of knowledge, an history which informs us that every advance in knowledge has been based on what is/was already understood. So, all understandings stand waiting to be understood. The insights of robotics, artificial intelligence, linguistics, medical narratives, neurology (including neural networks and maps), quantum physics, phonemes and genes, pharmacology, philosophy, religion (including Taoism and Buddhism, literature, psychology, biochemistry, anthropology, human ecology, among others, await out efforts.

6. Conclusion

In the presentation, I tried to touch on a number of problem areas – existential conditions, divergent philosophical outlooks, current structure and orientation to research, the history of that orientation, and the challenge as well as promise of the life sciences. There is an overarching framework which informs my own positions and one with which, I hope, you will wrestle.

---

12 Gary Gutting (ed), *The Cambridge Companion of Foucault* (Cambridge: Cambridge University Press, 2003), p. 22.

13 Michael Oakeshott, *On Human Conduct* (Oxford: Clarendon Press, 1975) pp. 1–32.

The universe and the earth within it seems to embody sources of fear but more so sources of grandeur and inspiration – a grandeur in which everything enjoys an identity that is in alignment with and seemingly signals to other things. Indeed, it appears that everything declares itself to other things, entering into innumerable relationships. The capacity for relatedness, therefore, is central to all things, and any approach to research and learning that limits the opportunities to recognize and elaborate these relationships and their properties is limited and limiting. Consciousness itself appears to be unitary – linked to a single "integrated perception that nonetheless changes endlessly, shifting from second to second through a never ending variety of states. [It] seems like a movie in the head, the pictures integrated not only with sounds but with smells, touches, tastes, emotion, feelings, thoughts, all bound up in a network called a sense of self, anchoring our own whole being and its experiences to our body."[14]

The life sciences have evolved in an era that appear to be most auspicious – not only does it encompass broad and deep disciplinary boundaries, as before indicated, but it is itself but imbedded in a new tradition (and must adopt itself thereto) of global trans-cultural and intercultural sharing, where the non-sequential thinking of Taoism and Buddhism meets with the largely more sequential approaches of the West. It has also evolved at a time when, because of the frustrations with existing disciplinary limits, we are urged to move beyond those limits, in part by transdisciplinarity. Perhaps most important – and this, I think, will come to define research in the life sciences – is the convergence on and increasing comfort with imaginings and emerging practices that emphasize networks, circuits, fluid movements between quantitative and qualitative analyses, pathways, cortical processes and routs, joint government-academic cooperation, the complementarity of Neils Bohr, the uncertainty principle of Heisenberg, the productivity-re-productivity of Buddhism, and the unity of opposites, by Taoism. In the latter regard, I would like to leave you with a quote from Lao Tse who, 2.500 years ago, wrote: "Only because of what is not there, it is possible that there is what is there; and only because of what is there, is it possible that there is not what is not there."

Where we are or are not are all understandings waiting to be understood – contingent holding places, for new departures in the life sciences and, through transdisciplinarity's rich collaborations, across universities, nations, and continents.

14 Lane, op. cit., p. 237. See, also, Thomas Berry, *The Great Work: Our Way Into the Future* (New York: Bell Tower, 1999); and Stanislas Dehaene, *Reading in the Brain: The Science and Evolution of Human Invention* (New York: Viking, 2009).

Healthcare Systems Research – Ethics and Reality

Heidrun Sturm

In the context of research in the life sciences field, health systems research has a special role. This article tries to make a link between biomedical and lifescience research, on the one hand, and healthcare systems research, on the other. In preparing this topic it became obvious that only a look at the overall healthcare system will provide clarity about the roles and contribution of the difference research fields.

In a second step, the contributions of hc systems research in general and, more specifically, in relation to cooperation structures are discussed. This may help researchers and healthcare industry to clarify strategic goals and their respective ethical and social implications.

1. The Healthcare Arena

A healthcare system can be defined as the means by which societies provide support for citizens to maintain their good health.[1] Therefore healthcare (HC) systems are complex organizations which need to continually adapt to external changes and internal challenges. In this sense, the complex system of healthcare delivery can be approached as a learning system, thriving to continuously improve its performance and minimize failure. For analytical reasons an adapted framework according to the Plan-Do-Check-Act – the "PDCA" – cycle of continuous quality improvement theories can therefore be helpful for understanding the various parts of the system: roughly speaking, "Plan" implies defining goals, and adjusting the system according to needs. "Do" refers to the provision of healthcare itself. The system's "framework" is defined by politics and stakeholders. "Check" refers to evaluation of the outcomes, an original task of healthcare systems research.

---

1 Here we consider health *care* systems in the stricter sense, concepts of health systems are wider – compare the WHO-Definition of Health Systems: "A health system consists of all organizations, people and actions whose primary intent is to promote, restore or maintain health. This includes efforts to influence determinants of health as well as more direct health-improving activities. A health system is therefore more than the pyramid of publicly owned facilities that deliver personal health services. It includes, for example, a mother caring for a sick child at home; private providers; behaviour change programmes; vector-control campaigns; health insurance organizations; occupational health and safety legislation. It includes inter-sectoral action by health staff, for example, encouraging the ministry of education to promote female education, a well known determinant of better health."

Fig. 1: The Healthcare-Arena

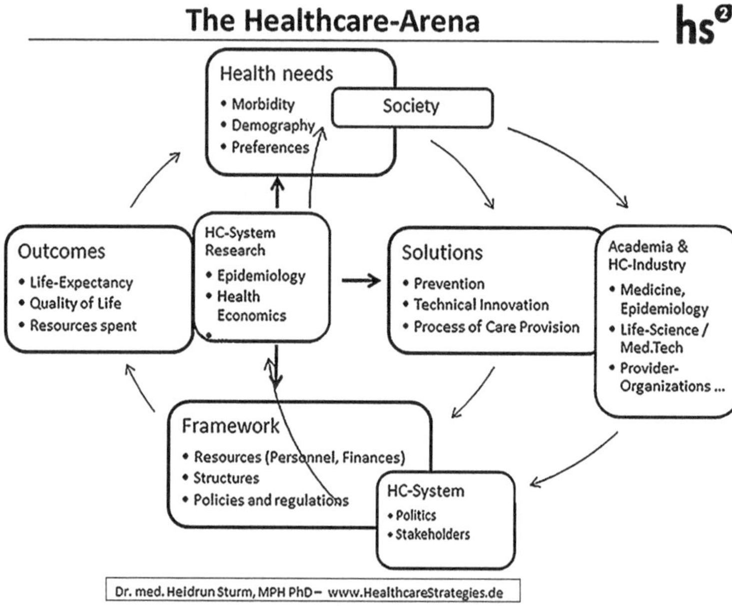

Source: HS2

The goal of healthcare systems – to maintain or improve the healthcare status of a given population – can be divided based on the WHO criteria to measure a system's performance into at least three parts: firstly, improving health (effectiveness); secondly, responsiveness to peoples' expectations and needs; and, thirdly, using the financial means in a fair way.

2. Needs and Preferences

Thus, in order to be responsive, the *health needs* of a society have to be assessed and defined: demographic and epidemiological data provide evidence about the prevalence and development of diseases. These results help to focus on relevant research topics: in which areas will increasing numbers of patients be in need of therapy? Which are attractive fields to invest in? Are there underserved indications or patient groups (like rare diseases, orphan drugs)? Can cost increases or shortages in defined areas be foreseeable?

Besides medical needs, structural requirements like the manpower of healthcare provision also have to be considered. Morbidity data combined with regional healthcare delivery data facilitate planning and resource allocation. As an example, the upcoming shortages in primary care, especially in rural regions, which many societies are confronted with (US as well as Germany, Switzerland and many more) can be anticipated.

In addition to the sheer numbers, planning has to take culturally defined *preferences* into account. Every society needs to define preferences with respect to financing and provision of care based on its cultural values. Questions about the range of services covered by public funds or how to distribute scarce resources such as organ donation are other important issues. The answers to these questions are not homogenous. This is already reflected in the formulations of countries' individual goals towards healthcare, where usually the three WHO main criteria are all included but weighed differently. For instance the UK aims at "comprehensive, high-quality medical care to all citizens on a basis of meeting professionally judged medical needs and without financial barriers to access". Also the US aims at universal access to high-quality, comprehensive, cost-effective health care. Yet the current American health care reform debate provides a very clear example for the weight attributed to universal access: private freedom of choice (like the freedom of having health insurance) in wide parts of American society is considered much more important than in most European cultures, where social security systems are considered a constitutional right and, as such, more important than personal freedom.

Figure 2 shows the percentage of private expenditure, which mainly consists of private insurance policies and out-of pocket payments of patients. In the US, where half of spending comes from private funds, private insurance plays the major role.

Fig. 2: Private healthcare spending in 2009

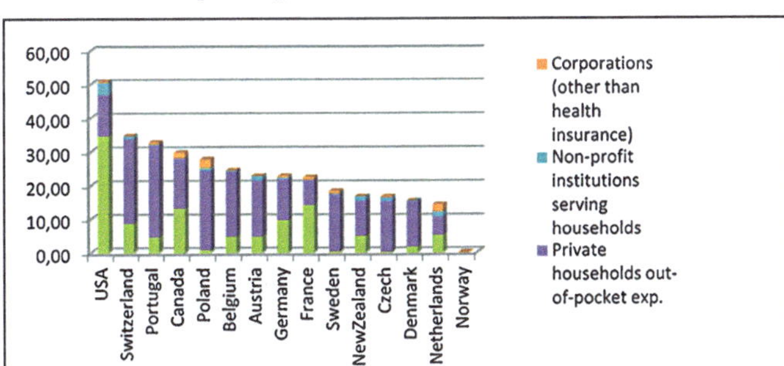

Source: OECD Health data 2012

3. Solutions

Needs are then addressed within the health*care* system by medical or organizational *solutions* such as technological innovations for diagnosis and therapy.

Also innovative care structures contribute to solving the challenges of healthcare delivery. Considering the aging societies, effective prevention is another increasingly important part, also within the curative healthcare system, of addressing the changing needs of healthcare.

Solutions within the care system are mainly developed and provided by academia and the healthcare industry in a wider sense. (Bio-) medical research and life sciences contribute to innovative therapies and technical solutions, while providers, healthcare researchers and other organizations like health insurers develop new care provision structures and prevention strategies.

4. Framework

Finally, all theoretical solutions are applied within a local *context*, where available resources in terms of personal or funds are equally as important as the underlying legal and organizational structure. This framework is predominantly defined by the underlying healthcare system. Whereas medical approaches like diagnostics and therapies are increasingly based on evidence and should therefore generally not be affected by the surrounding system, the process and organization of care by contrast can only work when it matches the local conditions.

Health care systems can be divided into tax-based and social security-based systems. In Fig. 3 it becomes clear that in systems which are mainly publicly funded, both variants are present: the Scandinavian countries and the UK are mainly tax financed, whereas the Netherlands, France and Germany, for instance, are typical social insurance funded systems.

Fig. 3: Public healthcare spending (social security and governmental expenditure) 2009

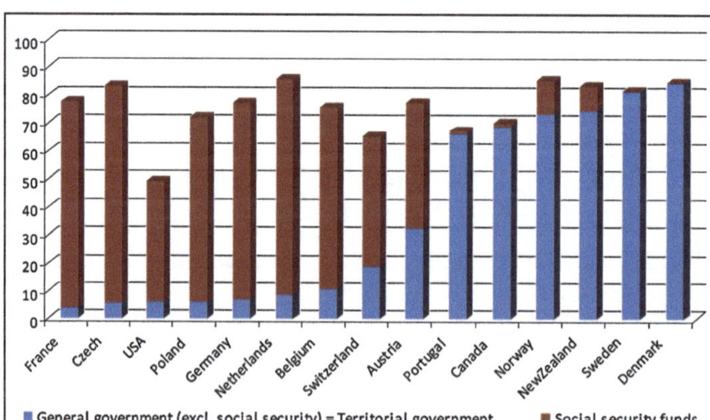

Source: OECD Health data 2012

5. Outcomes

The results of all joint efforts within the healthcare system produce outcomes in terms of improved health or quality of life. These outcomes should be routinely assessed in order to control the effectiveness and efficiency of the resources spent. Hard endpoints such as survival rates are important, as is quality of life or spent resources. But also process parameters such as adherence to guidelines, numbers of procedures performed etc. may help to point to inefficiencies and to detect opportunities for improvement. Only with such data is it possible to estimate if the provided care system is indeed fulfilling its premises, if resources are spent in a useful way.

Often the importance of this part is not sufficiently acknowledged. Evaluations are indispensible for adjusting and improving the HC system, taking social, political, economical and cultural aspects into account. The latter is especially important, as the overall goal with respect to producing health is often narrowed down to curative medicine.

6. Healthcare system research and research in life sciences

Coming back to the question of how HC research relates to research in life sciences, it becomes clear that while life sciences research provides concrete solutions, HC research provides theoretical input on practically all levels of the system in order to increase transparency and facilitate further development. This implies that HC research focuses on the system as a whole; it should provide the evidence base for all aspects of planning and decision-making – free of individual stakeholder interests. According to WHO, Health Policy and Systems Research seeks to understand and improve *how societies organize themselves* in achieving collective *health goals*. In order to be able to cover all different aspects of the complex health care system, research areas comprise a wide range of fields, such as epidemiology, as well as health economics, ethical issues or structural and organizational topics. Thus, HC research by definition needs to be interdisciplinary in terms of contributing professions: public health, medicine, economics, sociology, anthropology and political science, to name the most common professions involved in this field of research. In a recent call of the German Federal Ministry of Education and Research (BMBF) Healthcare research is defined as research describing and explaining healthcare provision and its contextual framework. Evaluations should target everyday conditions (effectiveness and efficiency) and derive new healthcare concepts based upon these evaluations. This includes different disciplines like rehabilitation, nursing, palliative care, as well as health economy and healthcare systems research.[2] The Agency for Healthcare Research and Quality, 2002 *defined* Health services research in 2002 as: [it] examines how people get access to health care, how much care costs, and what happens to patients as a result of this care. The main goals of health services research are to identify the most effective ways to organize, manage, finance, and deliver high quality care; reduce medical errors; and improve patient safety.[3]

7. Present challenges of HC systems and contribution of HC research

At present, the demographic development and the resulting consequences for care provision and funding are dominant topics discussed in virtually all industrialized nations. As a consequence of double aging – meaning the relative and absolute increase of elderly in the society - the spectrum of diseases is changing. Morbidity is moving towards chronic diseases due to longer life expectancies and better therapies. With people living longer and having *chronic diseases, multimorbid patients* are becoming a major group of patients seen in outpatient practices. In primary care alone, the number of patients with chronic diseases doubled, the proportion of patients with 3 and more diagnoses tripled between 1985 and 2005.[4] This has different implications:

Firstly, as the treatment of these diseases induces increasing health care costs, again driven by innovation and potentiated by the aging society (yearly health care expenditure increases with age), urgent strategies are needed to prevent these diseases in the first place (primary prevention), but also to prevent their progress (secondary and tertiary prevention, which are mostly relevant to the curative healthcare system). This implies that potentially the relation between curative and preventive medicine needs to be revised. Questions related to that are:

---

2 Bekanntmachung des Bundesministeriums für Bildung und Forschung von Richtlinien zur Förderung von Studien in der Versorgungsforschung Vom 15. August 2012. Versorgungsforschung wird hier verstanden als die Wissenschaft, die Kranken- und Gesundheitsversorgung und ihre Rahmenbedingungen beschreibt und erklärt, unter Alltagsbedingungen evaluiert und darauf aufbauend Versorgungskonzepte entwickelt. Dies schließt unterschiedliche Disziplinen ein, wie z. B. Rehabilitationswissenschaften, Pflegeforschung, Forschung zur allgemeinmedizinischen Versorgung und Palliativmedizin sowie Gesundheitsökonomie und Gesundheitssystemforschung.

3 Dep. of Health Services, School of Public health, Univ. of Washington (http://depts.washington.edu/hserv/hs-research-definitions).

4 Uijen & v.d. Lisdonk (2008) Multimorbidity in primary care: prevalence and trend over the last 20 years. Eur J Gen Pract 14(1): 28–32.

- how will available funds be distributed between the two areas;
- how much and what aspects of care will be financed by society and its social security schemes, and what will have to come from private funds. Most likely, the so-called second healthcare market with its very different market structures will play an increasing role in the future.

Secondly, chronically ill patients need integrated care structures: they are long-term patients and their care most importantly has to focus on preventing progression of the disease. These patients typically need care from different care sectors and disciplines. Therefore, coordination of care is vital to prevent the waste of resources and improve outcomes.

Thirdly, multi-morbid patients pose additional - especially medical - challenges: co-existing diagnoses require various therapies, resulting in highly complex treatment regimes. Not only are they difficult to follow by the mainly elderly patients, but they also bear the risk of multiple drug-interactions. In a frequently cited case study by Boyd,[5] *a 79 year old lady with five diagnoses (osteoporosis, arthritis, Diabetes Mellitus type 2, Hypertension, COPD) is presented. Her guideline-based treatment regimen imposed nine drug-drug interactions, two drug-disease interactions and eight drug-food interactions. The lady had to follow 20 recommendations for diet and life-style and had to take 19 single drug doses in a five times daily complex regimen. In order to prevent exacerbation, frequent control visits to the physician practice were necessary. The monthly costs were immense.* It becomes obvious that this situation bears multiple risks where things can go wrong on the patient side (compliance) as well as on the provider side (coordination). Measures and structures addressing these problems are not sufficiently available at this point. Here HC research demonstrates deficiencies in care which, together with developments in demand, will help industry to define promising areas for research and development. For this common type of patient, drug blistering, IT communication platforms or telehealth seem to be examples of promising approaches.

Another aspect to this is the lack of evidence concerning multi-morbidity. Effectiveness studies based on randomized controlled trials (RCT) mostly include very selective patient groups with a defined disease. Also, guidelines deal with individual diseases and often result in contradictory recommendations in these situations. Thus, evidence-based medicine also needs to widen its research focus and address therapies for multi-morbid elderly patients, relevant to pharmaceutical industry as well as clinicians.

---

5 Boyd et al., JAMA 2005. 294: 720 ff.

8. Political and strategic approaches

These challenges are approached in different countries with various, yet similar approaches. They include programs to adjust financial incentives for providers (more financial responsibility for regional structures, pay for performance etc), improve patient literacy, coordination of care approaches. In this article some challenges in the "SPANNUNGSFELD" of politics, evidence and corporate interests will be discussed based on the German example:

- *Disease Management Programs and Politics*. In Germany, as in other countries, disease management programs (DMPs) were seen as potential solutions to improve care and contain expenditure for common chronic diseases. Contrary to many other countries however, Germany linked DMP programs to allocation of health insurance funds (Risikostrukturausgleich) and at the same time implemented a huge administrative body around them. Sickness funds with a high share of DMP participants received higher compensation from the risk structure compensation scheme and therefore had an incentive to sign up patients, but no incentive to improve care.[6] In addition, the program was implemented nationwide at the same time therefore making any evaluation using control groups impossible[7]. This whole process implies that, rather than being focused on medical effectiveness based on HC research evidence, implementation of DMPs was pushed due to political and corporatist interests. Now, after ten years, in circles of political decision-makers the program is being presented as a big success, mainly based on increasing numbers of participants[8] lacking clear results on effects.
- *Increased competition, innovative structures and the role of incentives*. In order to facilitate implementation of intersectoral and interdisciplinary care structures, increased market orientation is a commonly applied recipe: in the German setting of hitherto strictly restricted ambulatory care, provision in conjunction with the provider organizations (Kassenärztliche Vereinigung) (see Fig. 4) in recent years has offered increasing possibilities for *selective contracting*: since then, physicians or provider networks have been able to develop individual care

6 Busse R: Disease management programs in Germany's statutory health insurance system. Health Aff 2004, 23: 56–67.

7 Schäfer et al., BMC Health Services Research 2010, 10:351. http://www.biomedcentral.com/1472-6963/10/351.

8 Veranstaltung „10 Jahre DMP", Jan. 2012 Berlin. Versorgungsforschung Bundesversicherungsamt.

models and alternative reimbursement modes and negotiate directly with sickness funds avoiding physician associations'[9] strict rules.

Fig. 4: Health Care's Fragmented Payment Streams

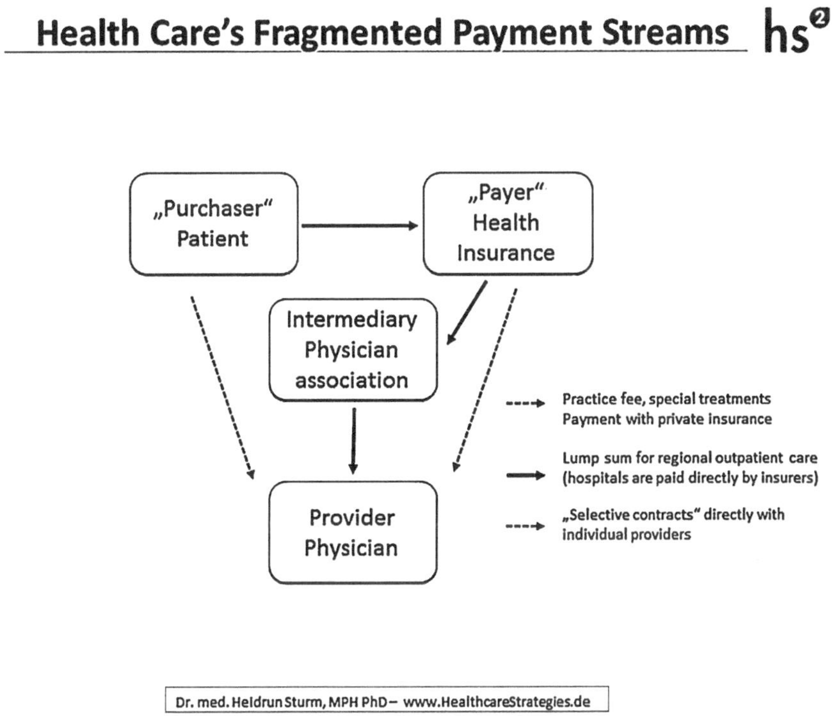

Source: HS2

In this context, gate-keeping models for primary care were implemented, and similar so-called "structural contracts" (Strukturverträge) were developed for certain specialties or indications like orthopedics, gastroenterologists or neurologists. A special focus was put on *integrated care contracts*,[10] which were intended to overcome the strong German segregation of inpatient and outpatient care and other sectors such as rehabilitation. This enabled especially networks of providers

9 Kassenärztliche Vereinigung.
10 §140 ff SGB V.

to set up care models, where different medical specialties and other caregivers including hospitals work together to care for a given patient population. Usually these organizations negotiate budgets with individual funds, based on costs of comparable populations. As a result, the provider network is free to use any generated surplus e.g. for reinvestments within their structures. Theoretically this creates incentives to cooperate effectively within the network and – if the system is setup for a longer period – at the same time aim to improve the health status of the served population. Evaluations of the cited models point towards these results at least during the first years:[11]

- *Market forces within a regulated system.* In Germany this system is complicated by the number of sickness funds, which makes it difficult to offer programs for a regional population rather than a given group of insured persons. The latter varies from provider to provider and from region to region, usually not covering more than 20 to 30% of a practice population. So the goal of these provider organizations must be to convince multiple sickness funds of the concept. However, despite the promising evaluations it appears to be difficult to multiply successful concepts such as the "Gesundes Kinzigtal". One of the reasons might be that sickness funds are not only interested in saving, but at the same time are competing by means of "outstanding" offers for their insurees and therefore mostly are not interested in offering similar programs to their competitors. This is especially true if various selective contract versions like gatekeeping are available in the same region. - Also, for providers it is difficult to work with countless care models depending on the insurance status of a patient. Therefore it seems likely that financial incentives of individual contracts will be weakened by multiple co-existing contracts. On the other hand, quality effects and cost efficiency savings of good contracts may spill over to patients outside those contracts.
- *Industry as a supporter of providers.* Integrated care contracts were originally targeted at providers and sickness funds only. Since 2011 pharmaceutical or medtech companies can also be partners in such contracts.[12] Thus, in 2011 many

11 An example provides the regional approach of the "Gesundes Kinzigtal" in the Black Forest or the UGOM physician net in Bavaria, where provider networks offer the complete range of care for their participating patients. See also: Hildebrandt, H.; Hermann, C.; Knittel, R.; et.al. (2010): Gesundes Kinzigtal Integrated Care: Improving Population Health by a Shared Health Gain Approach and a Shared Savings Contract. In: International Journal of Integrated Care 10, S. 1–14 (http://www.ijic.org).

12 AMNOG: Arzneimittelneuordnungsgesetz.

companies started out to seek partners and design programs which promised to improve care, are attractive for providers in some way and contain costs for insurance companies at the same time. In order to be attractive to industry as well, those programs should usually coincide with major indications of their activities.

Also, before 2011 industry could cooperate with providers by offering support of various kinds. According to Janssen Cilag (a daughter of Johnson & Johnson), industry's role can be based on its medical and scientific knowledge based on its products, on knowledge in management of complex organization and process management, as well as development of business models. Industry can also provide financial support.[13] Common activities are, for instance, administrative support for networks (like IT structures), but also include support for pathway development or group-specific positive lists.

Lately these activities have been under close observation by media and authorities. Concerns focus on the independence of physician prescribing and in 2011 the news magazine "Der Spiegel" titled: "Favors on Prescription: How Physician Networks and Pharmaceutical Industry Support Each Other".[14] The article reported about a physician network whose prescribing pattern suggested that instead of allowing generic replacement (aut idem), defined products of the supporting pharmaceutical company were prescribed.

But also the less straight forward construct of Janssen-Cilag is under media supervision: the management organization I3G managing a schizophrenia network in lower Saxony is owned by Janssen-Cilag,[15] thereby bringing the pharmaceutical industry into integrated care contracts even before the new legislation in 2011 was passed. Like in the examples above, the integrated care model seeks to be more cost efficient – especially by avoiding hospitalizations. The virtual savings (compared to care as usual) are split between the sickness fund and I3G. The sickness funds hope for additional profits through optimized coding by network physicians and therefore improved allocation of funds from the risk structure compensation

---

13 Integration, Hybridisierung und Entstehung von Systemanbietern. Die Zukunft personenbezogener Dienstleistungen im Gesundheitswesen. Dr. K. Suweleck, I3G GmbH. IGES: Zukunft personenbezogener Dienstleistungen 2012. Leistungen_2012_Suwelack_Integration,Hybridi sierungundEntstehungvonSystemanbietern_ger.pdf).

14 Der Spiegel, 2/4/2011.

15 Janssen-Cilag is owned by the US-company Johnson & Johnson and produces amongst other various neuroleptics (Invega®, Risperdal® and the depot-neuroleptic Consta®).

scheme.[16] Also, in this integrated care concept suspicions were raised towards potential influence of the industry with respect to provider independence, and intransparency of financial flows was also criticized. Among the critics are physician organizations, as well as patient groups. They fear financial interests will dominate quality of care.[17] Despite the slow sign-up process of patients, the sickness fund is now opening up the concept to more regions in lower Saxony, as well as hospitals. The first evaluations are still awaited.

The first real "three-sided" integrated care contract, which was established in fall 2011, has since become quietly implemented. A sickness fund, a pharmaceutical company and on the provider side an academic center and a specialist center[18] are partners in the contract. Physicians work in a network including pediatricians in private practice; industry is providing a telecommunications platform, which also includes patients; the sickness fund provides the contractual framework with reimbursement schemes for participating providers. Here the set-up of the co-operation might be more clear-cut and therefore less prone to subtle influences.

Since then hardly any additional contracts have been published in the media. Obviously all those constructs which include the participation of industry are very difficult to mediate to patients and public, as implicitly interests of insurance companies, as well as industry, ought to put economics first before focusing on patient interests. Therefore criticism is mainly focused on the independence of physicians' therapeutic freedom. While this is a very justifiable concern, it is often neglected that influence on physicians' therapeutic decisions does not start with contracts like this, but is already compromised by their own financial constraints and interest in securing their own income. It is also restricted by sickness funds' drug rebate contracts, which increasingly determine the generic substance a patient can be reimbursed for.

Additionally, three sided contracts between payers, providers and industry have proven to be difficult in terms of creating clear win-win-win situations. The biggest obstacle is to produce savings outgrowing initial invest within a one or two-year period. These time-frames are, however, still common ROI periods of sickness funds. Recognizing this shortcoming, German politics has reacted and offers sickness funds alternative financing modes, such as special credits for such contracts. It remains to be seen if these measures will improve uptake of innovative integrated care structures.

---

16 Ärzte Zeitung, 15.08.2011.

17 Financial Times Deutschland, 23.8.2012: Schizophreni: Projekt mit psychisch Kranken gescheitert.

18 Epilepsiezentrum Raisdorf für Kinder- und Jugendliche.

To Create and Control Life – A Critical View on Synthetic Biology

Rainer Paslack

As has often been said, synthetic biology is one, and perhaps the most ambitious research program to apply nanobiological methods in the fields of biology, medicine, environmental technologies, and some others. But first of all: synthetic biology is the attempt to create and to control life. Indeed, this sounds very ambitious! And it will be one of the purposes of this article to examine whether – and in what respect – this aspiration is justified.

1. Synthetic biology – Aims and methods

What is "synthetic biology"? There are various definitions covering most aspects of this new research field. As an example, among many, the European Commission gives the following definition:

> "Synthetic biology is the engineering of biology: the synthesis of complex, biologically based systems which display functions that do not exist in nature. This engineering perspective may be applied at all levels of the hierarchy of biological structures – from individual molecules to whole cells, tissues and organisms. In essence, synthetic biology will enable the design of 'biological systems' in a rational and systematic way." (European Commission 2005)

In other words: the ambitious tasks of synthetic biology are "the design and construction of new biological parts, devices, and systems, and the re-design of existing, natural biological systems for useful purposes" (Endy 2005). And Bhutkar defines – in analogy to the construction of a computer – as the "primary objective" of synthetic biology that it "creates a programmable microorganism from scratch" (Bhutkar 2005: 19 f.) Moreover, it should be increasingly possible to "program living organisms in the same way a computer scientist can program a computer" (The BioBricks Foundation 2010). From this perspective the approach of synthetic biology seems to be guided by the paradigm of computer technology, so that "artificial life" can be seen as an analogy to "artificial intelligence".[1] It is obvious that

1 And, consequently, like computer software, "biological software" as a product of human imagination also has to be considered as copyrightable and patentable (cf. Torrance 2010: 646).

synthetic biology is primarily led and driven by technological aims, more than by interests of mainly biological background.

Historically, the term "synthetic biology" is an old one, already coined (as "biologie synthétique") by Stéphane Leduc in 1912 (Leduc 1912). A more modern version was given by the geneticist Waclaw Szybalski in 1974, who wrote: "We will [...] devise new control elements and add these new modules to the existing genomes or build up wholly new genomes. This would be a field with the unlimited expansion potential and hardly any limitations to building 'new better control circuits' and [...] finally other 'synthetic organisms' [...]." (Szybalski 1974)

Currently – and based on the success of genetic engineering – synthetic biology is a rapidly developing new field of research that aims at extending the abilities of genetic engineering.[2] The research is often described as "creating life"[3], a description that is bound to catch public and political attention; it is therefore more an often-used label in public discourse than a scientific self-description.
There are two main approaches in synthetic biology research:

(1) The top-down approach
(2) The bottom-up approach

(1) According to the so called *top-down (or "reducing") approach*, relevant areas within this field include the attempt to reduce the genome of a microorganism to such a degree that the organism is capable of living, but, beyond that, does not express any specific abilities. The idea is to equip such a "platform" organism with whatever genetically encodable ability one may conceive of. This approach is therefore also often called "reducing approach", and has been used successfully, for example, to reduce the size of the *E. coli* genome. In other words: the scientist starts with a whole microorganism, or its whole genome, and reduces it in such a manner that one can study, for instance, certain pathways of signaling, or the process of gene expression, or other gene-based functions.
(2) In opting for the so called *bottom-up approach* or "*de novo* synthesis approach" researchers try to build protocells from the bottom up, using smaller, separate molecules. Beyond this, the result could be organisms whose genetic

---

2 Anyhow, the first synthetic biology conference was already held at MIT in 2004. Most research is still university-based, but there are also several private companies, like Synthetic Genomics or Amyris Biotechnologies (cf. Motavalli 2010).

3 As the attempt to "engineer" and "program" living cells and their genomes, synthetic biology allowed "a conceptual break with the previous, largely descriptive, molecular biology, and ushered in a new paradigm of genetic manipulation at the molecular level." (Torrance 2010: 634)

make-up is based on parts not found in naturally occurring organisms. In other words: the scientist starts with so called "biobricks", e.g. single genes, and produces the whole entity (an organism) step by step. Some examples: The Craig Venter Institute in Rockville (USA), whose scientists are working on the *de novo* synthesis of the *Mycoplasma genitalium* genome is a driving force within this approach. The Craig Venter Institute has also shown that genomes of bacteria can be exchanged, which is one prerequisite for introducing the synthetic genome, once it has been built, into a host bacterium. The synthesis or assembly of the synthetic genome thus remains the main technical challenge. This has, so far, only worked for viral genomes which are much smaller than the genome of any minimal organism. Among the first synthetically viral genomes, that of polio virus and bacteriophage X174 are to be found.

Still more ambitious are projects which include the *in vitro* creation of alternative genetic alphabets, expansion of the genetic code to unnatural amino acids, or RNA syntheses that read quadruplet instead of triplet codons.

Possible goals of research in the field of synthetic biology are

- the production of new drugs (e.g. *artemisin* against malaria[4]),
- the creation of micro-organisms for medical purposes (e.g. for the treatment of cancer), or as "vectors" for gene pharmaceuticals,
- the replacement of dead human tissues,
- the production of new biological based fuels (e.g. ethanol),
- the production of biosensors and biological chips.

The list above contains only the most interesting possible applications of synthetic biology; it would be easy to add more.

Overall we can characterize synthetic biology research in a twofold manner:[5]

1. The first is *interdisciplinarity* (on a more scientific level): synthetic biology often combines approaches from various disciplines, such as biotechnology,

4 See e.g. Martin et al. 2003.

5 Similar Rutz 2009: S14 f. Compare Endy, who differentiates between "three general principles" guiding synthetic biology: (1) "standardization" in the sense of the definition and characterization of the basic biological components, (2) "decoupling" as decomposing larger tasks into smaller tasks, and (3) "abstraction" as (a) "breaking a biological engineering problem into hierarchical levels of complexity, and (b) "redesigning the basic components of engineered biological systems to simplify the construction and deconstruction of such systems". (Endy 2005: 450 ff.)

chemistry, nanotechnology, computing and engineering. So it combines the systematic analysis of intracellular genetic, metabolic, and signaling processes with the synthesis and rebuilding of novel single cell organisms, using engineering and IT principles.
2. The second is *complexity* (on a more technical level): synthetic biology often involves many 'parts' or components that work together to achieve a certain effect. In consequence of this, the complex structure of a synthetic or artificial organism could generate non-linear effects which are difficult to predict (especially if the organism interacts with an environment which is also of high complexity).

In attempting to alter genetic structures of organisms, synthetic biology is closely related to genetic engineering. First of all, like genetic engineering synthetic biology is a technological endeavor, driven by technological interests rather than scientific purposes. Both genetic engineering and synthetic biology are attempts to modify or create organisms, in the sense of "machines" producing interesting substances or performing desirable operations in a *controlled* manner. Therefore it is not intended that the new organisms (especially if they have been released into the environment) behave like natural organisms with the ability to evolve in an unpredictable manner. From this perspective genetic engineering and synthetic biology are both examples of a "technological reductionism" with regard to living beings. Especially this aspect will be the focus in the third part of this article.

Nonetheless, the two fields of research/technology differ in some respects:

- Synthetic biology is based on a systematic analysis of intracellular processes;
- it aims at altering metabolic pathways, signaling processes, and DNA sequences on a large scale; and
- it does so by using engineering principles of modularization and standardization.

Hitherto we can summarize: in order to account for the ethical, social, and epistemological relevance of synthetic biology, these differences have to be systematically explored and evaluated in the future. In the following we will only sketch some important aspects of the complex questions involved in synthetic biology.

As a result of extended insights into the functions and modes of action of the genome, our abilities to manipulate and replace genome sequences have increased. What was initially a simple gene technology in the second half of the 20th century (with the purpose of producing slight genetic *modifications*) is currently evolving (under the term "synthetic biology") into a technique that permits single-cell organisms to be tailor-made according to our own wishes, and thus *creating completely new forms of life*. The climax of this development, at least for the time being, was the patent application submitted in 2006 by a group of USA researchers around Craig Venter for a minimal bacterial organism genetically engineered in the laboratory as a "Mycoplasma laboratorium". Alarming news like this is the reason why synthetic biology attracts the attention of the media (and the public) to such a great extent.

But most synthetic biology projects are still at a *visionary* stage, and it is therefore difficult to estimate how realistic the hopes of the "makers" really are.

Now I will move on to the second part of my investigation dealing with some ethical and epistemological questions involved in synthetic biology.[6]

2. Ethical implications of synthetic biology

In the following, some ethically relevant questions will be addressed which concern a more philosophical or anthropological level of ethical debate. I would like to restrict myself only to ethical questions which are *specific* regarding synthetic biology. The ethical and epistemological implications of synthetic biology confront us with questions like the following:

- What does "life" mean when the biochemical synthesization of life is involved?
- Is an artificial cell a living being or an artifact? Or: in what respect can a living cell be artificial?
- Could synthetic biology cause effects on the inherent values in non-human living beings?
- Could synthetic biology open a new perspective on nature?

It is hardly surprising that this technology provokes a heated response. Now, the term 'life' denotes not merely a *descriptive* but also a *normative* concept. Its meaning comprehends more than metabolism and self-organization, or the ability to grow or to evolve. Vitality is also at the core of what implies worthiness of protection, not least when living things display evidence of pain perception and simple forms of awareness; moreover, vitality is often related to some properties like

---

6 I will not discuss the legal implications of synthetic biology related to possible risks like "biosafety" and "biosecurity", because these aspects have to be focused on from a more legal or regulative perspective.

unpredictability and individualism.[7] Therefore, finally, we often associate vitality with an *inherent* or *intrinsic* right to exist.

So, creation of life in the laboratory is considered one of the literary and cultural codes for the goals of modern humanity (like it was in the romantic myth of the Frankenstein monster). This is especially so since one of the objectives of this subdiscipline of biology is precisely to produce microbial forms of life with properties that do *not* occur in nature. Faced with the visions of synthetic biologists, old ethical and philosophical issues, expectations and public anxieties regain their acuteness. The fact of the matter is that nanobiotechnology, as well as synthetic biology, go one step further than previous techniques in their ambitious approaches and aims, since novel types of "hybrids" at the interface of living and non-living matter will be constructed, which call for a revision of the difference between living organisms and machines, and raise ethical questions about the value of such organisms, as well as questions related to their synthesis.

Most of us intuitively make a difference between living and non-living matter. This difference is based on cultural, social, philosophical, religious and ethical criteria of life. And most people agree that human life has a certain value – but what about life of other organisms? Do they have an intrinsic value, too, in the sense of the biocentric position? Or do they only have an "instrumental value" related to their usefulness to human beings, as the supporters of the so-called "anthropocentric position" argue?

In addition, from an ethical and philosophical perspective, the shift from *manipulation* in individual cases to *creation* is significant and pertinent. "Life from the lab" has become a catchword to attract attention from the public.

Now, an organism, for example, which has been synthesized by the minimal genome approach, would clearly be living according to all criteria which are valid for natural organisms. But one might consider that there could be another type or degree of life than in the case of natural organisms (regarding such properties as metabolism, self-organization or evolution). This question will become particularly interesting should an artificial type of nucleic acids or an unnatural genetic code be used one day for the synthetic genome. And it is not clear whether they would have the same value as natural organisms. It may be that we – as their creators – would be more responsible for organisms we have created than for organisms occurring naturally in the environment.

Therefore, firstly, it seems of high ethical importance to analyze descriptions and concepts related to the entities synthetic biology is about to produce. A term

7 So one could ask: "Is biology impenetrably complex, unmanageably complicated, and essentialiy unpredictable, or can biological systems and their components be understood, manipulated, and controlled to an extent sufficient to synthesize artificial versions?" (Torrance 2010: 637)

like "artificial cell" may serve as a case in point: is an artificial cell a living being or an artifact? In what respect can a living cell be artificial? It may be that there is no contradiction and that a cell can be both: living *and* artificial. But what does it mean to call a living cell "artificial"? As can be seen, a series of conceptual, ontological, and anthropological questions and reflections arise from the hybrid concepts of synthetic biology research. Some of these questions are related to the creation of single- and multi-cell organisms. Taking it for granted that synthetic biology can create new organisms, it needs to be examined whether this characteristic is bound up with ethically significant implications – especially if the difference between artifacts and living entities may become blurred. Even among biologists it is not clear what "life" exactly means – descriptively as well as normatively. Accordingly, it is still questionable whether it is justified that critics of synthetic biology reproach synthetic biologists for "playing god" (cf. Schummer 2011).

A second point of ethical relevance is the question: might synthetic biology have implications for the *inherent values in living beings*, on condition that also non-human organisms *have* such values? – The term "inherent (or intrinsic) value" implies that an organism is worthy of protection for its own sake. Now, although virtually no ethical theory counts single-cell organisms among living beings inherently worthy of protection, it will be important to determine at what point in its development life begins to be seen as inherently valuable according to different ethical premises. Such a reflection will help us to understand if, and in what respect, the creation of an organism in general might come into conflict with its inherent value. Furthermore, it will help to identify criteria for deciding between organisms worthy of protection and those not worthy of protection.

The question of value inherent in living beings will be particularly pertinent in the case of synthetic multi-cell organisms. Although synthetic biology research is far from accomplishing this task, first steps have been taken to build multi-cellular compounds that might in the future be used to form organs. With regard to this research, it will be important to determine the relationship of synthetic biology and tissue engineering, and to assess its potential to synthesize *whole* organisms.

Now, if synthetic biological organisms are interpreted as artificial, machine-like entities without inherent value, it is to be asked whether this may in the long run change our attitude towards responsibility and respect for living beings. It might well be that we think it appropriate to show a higher or special responsibility for living beings we have created than for organisms already existing!

This consideration leads, thirdly, to the philosophical question: *Could synthetic biology open up a new perspective on nature?* It has been argued that the shift from manipulation to creation involves a shift in the way scientists and engineers approach the objects of their work. If we suppose that we always act against the background of a general conception of nature, which is a mixture of empirical fact and normative interpretation, the shift from manipulation to creation could have effects not only on our self-conception as "creators", but also on our perspective on nature as a whole. This shift in perspective can be ethically relevant in several respects. On the one hand, one may argue that the creational approach is a genuine expression of human creativity and of scientific inventiveness, and, as such, promises advances in the promotion of human welfare. With reference to 20th century environmental ethics, on the other hand, one may point to the assumption that unsolicited side effects of synthetic organisms may be just as unforeseeable.

All in all, synthetic biology is a field that promises significant enhancements to our understanding of nature. It may enable us to change nature in a novel and useful manner, but also in risky ways (with respect to biosafety and biosecurity). For an understanding of the ethical, societal and legal aspects of synthetic biology, mid- and long-term prospects for the field are needed.

This relates, among other things, to state-of-the-art technology in the field, its potential social and economic benefits, as well as its risks. However, due to its recent emergence and the highly visionary character of the discourse on synthetic biology, potential effects and implications of the field are difficult to *assess* in the sense of technology assessment and risk analysis. While there are quite a few starting points for benefit- and risk-focused assessments in the case of established fields of research, we have here to deal with extremely high uncertainties, and with a scarceness or even absence of evidence that are normally referred to in technology assessment projects. Regarding technology assessment, critics often tend to focus on distant-future prospects for the field, from which a highly speculative and strongly polarized debate can evolve.

3. Epistemological implications of synthetic biology

Finally, I would like – from an epistemological perspective – to examine the nature of synthetic biology more closely. Even assuming that synthetic biology will one day actually be able to fulfill its vision to create complete and novel (even multicellular) organisms that meet the needs of engineers for a well-functioning machine, then the question arises of what model SB actually follows. Questions arise such as:

1. Which paradigm is the leading one for synthetic biology?
   Now, the development of SB so far shows that apparently different epistemic paradigms are fighting for dominance: a biological and a mechanistic paradigm.

2. Is synthetic biology based on technological design still a biological discipline, for which "life" is a relevant factor?
   In fact, it seems to be that the technological interest of many synthetic biologists exceeds the limits of a strict biological understanding of life.
3. Is the view of production of "artificial life" a contradiction in itself?
   It is crucial, however, what the term "life", on the one hand, and the term "artificial", on the other, mean.

In answering these questions, I would like to dare – pointedly – the following theses:

1. Synthetic biology is currently changing from a biological to a more technological discipline, resulting in side-lining "life" as the initial idea.
2. In the ultimate vision of synthetic biology, "life" will be seen as a machine (an at most biomorphic technical system), and no longer as an organism.
3. From the perspective of synthetic biology, often the specific essence of life even appears to be "disturbing" because the new biosystems are designed for economically promising substances (i.e. drugs) and not for surviving in a stressful environment (for which the inherent adaptive flexibility of natural living beings is required).
4. On the other hand, the peculiar organization of organisms (their so-called "autopoiesis") offers important advantages (in contrast to machines of the classical type), which the bioengineers like to use for their technological purposes.
5. But it is doubtful whether it is possible to construct a system which can combine in itself both "designed technology" and "life".

For a better understanding of these rather strong theses I will contrast the basic characteristics of living beings with the basic requirements of a technical system. Perhaps the most important properties of life are:

(1.) *Self-organization and autopoiesis* (living beings are autonomous, self-producing and self-sustaining)
All living beings are self-organizing systems: they *organize* their internal structures (anatomy, metabolism, behavior etc.) by themselves. Moreover, they are also autopoietic systems, which *produce* themselves. "Autopoiesis" is a term coined by the Chilean neurobiologists Humberto Maturana and Francisco Varela (1980) to define the most essential property of each biological system. This means: all natural living beings. Autopoietic systems (like humans) are organized recursively, because the product of the functional

interaction of their components is precisely the organization that produces the components. The dynamics of such systems follow intrinsic purposes (in the sense of self-preservation). – By contrast: so-called "allopoietic systems" cannot produce or influence their own development. An example is an automatic production line for cars, in which the product has nothing to do with the production of related equipment. Such systems are the product of human actions based on "heteropoietic" extrinsic purposes.

It happens, however, that principally autopoietic systems are used *as* allopoietic systems (e.g. crops and livestock used by humans as food-producers).

(2.) *Non-triviality* (the predictability of behavior is restrained)

It was Heinz von Foerster (1999), a pioneer of cybernetics and system theory, who distinguished "trivial" from "non-trivial" machines (the term "machine" is here taken in a very broad meaning like the term "system"). A good example of a "trivial machine" is the computer: a particular keystroke always will bring a well-defined result on the monitor. If in the case of failure, either the software or hardware is defect, or the user does not understand the input-output relation. – By contrast: a system is a "non-trivial" machine (or system) if, when entering a specific input, the output is not exactly predictable. A funny example may illustrate this. Hitting a cow on her backside with a stick can cause very different reactions: the cow will go faster, or step aside, or hit back, or not do anything. – There are several possible sources of "non-triviality" in nature: the only limited predictability of behavior may be due to

(a) recursive output-input-linking (products or effects of the system or a part of it become new inputs into the same system),
(b) multifunctional mode of action of numerous components (e.g. genes which often have more than just one function),
(c) flexibility of individual behavior (e.g. adaptation towards changing conditions of the environment like coping with stress),
(d) spontaneous mutations or copying errors (DNA-replication), which raises evolutionary flexibility, and
(e) epigenetic alterations (variable gene expression after an environment-dependent "impression").

Sometimes even a third property will be proposed, but it could be that this trait is only (or mainly) important for those living beings which have a certain degree of awareness (consciousness) and self-perception:

(3.) *Internal perspectivity* (living beings are not completely definable from the perspective of an external observer like a biologist)

> This means: living beings follow intrinsic purposes, have intrinsic values and observe themselves in that they differentiate themselves from their environment. They are self-reproducing systems with a highly selective relationship towards their environment.[8] – Under the term "biosemiotics" there are many approaches that attempt to interpret "life" as a complex process of semiosis at all levels of a living system: from this perspective sign processes are understandable as meaningful and intrinsic for life as a whole, so that self-organized living systems are also governed by formal and final causality (not only by causality in the physical sense). It may be that this semiotic view of life marks a fundamental difference between living and non-living (technical) systems.[9]

Thus we can summarize that living systems (organisms) are self-organizing and autopoietic systems with a non-trivial behavior (and perhaps even an internal perspective with semiotic implications).

Regarding (non-living) *technical* systems of the classical type (whose behavior is fundamental trivial and therefore predictable), I would like to point out two basic characteristics which also paramount in the technological projects of synthetic biologists: *modularity* and *orthogonality*. Both properties of technical systems are strong connected, just as – in the case of organisms – autopoiesis and non-triviality are linked.

In the field of engineering and informatics the *modules* of a complex system are "*orthogonal*" if they act in a completely independent way, so that no unwanted interferences and malfunctions can occur. Orthogonality is required for reproducibility, control, operational reliability and safety of the system construct.[10] This engineering perspective is obviously also central to synthetic biology when the construction of a system is intended which has to act in a strictly controllable manner (e.g. to produce reliable drugs for medical purposes). – For the technological projects of synthetic biology, this means:

(1) biosystems have to be divided into functional subsystems (many modules), and
(2) undesirable interactions between the various subsystems have to be eliminated or at least limited without affecting the "viability" of the system.

---

8 But it is still controversial whether we can ascribe an internal perspective to organisms which do not have self-awareness in the manner of self-consciousness.

9 See Kull (1993); Sebeok (1999); Hoffmeyer (2008).

10 Compare Billerbeck and Panke (2012), and Schmidt (2012).

It is obvious that the non-trivial organization of organisms is not really compatible with the orthogonality of modularized trivial systems like machines of the classical type (e.g. a computer).[11] Compatibility could be achieved only if it were possible to construct a system that is both: autopoietic *and* trivial, because its structures – although autopoietic – are organized in a modular and orthogonal way. At present, this seems to be a paradox and apparently impossible – but the time may come.

4. Conclusions

Finally, concerning the epistemological status of synthetic biology, the following three conclusions can be drawn:

1. A definition of what "life" essentially is seems not to be necessary for the technological purposes of synthetic biology: It is sufficient that the entities that are usually considered to be alive can be modified or newly created in such a way that they fulfill the usual requirements for a complex technical artifact: in particular, programmability, controllability, predictability and safety. – In synthetic biology, the ultimate goal is the production of a trivial machine with a fixed output, which nevertheless exhibits features characteristic of a *non*-trivial machine like self-repair ability, self-replication, error-friendliness (robustness and resilience) and possible growth. But is this possible?
2. Therefore synthetic biology is an attempt to use heteropoietically constructed biosystems (although biosystems are naturally non-trivial beings) as trivial machines for human purposes. – The question is whether and to what extent this is possible! In other words: is it possible to produce systems with a "*trivialized autopoiesis*"?
3. Synthetic biology follows the paradigm of a "technological reductionism", although the "design" of the intended biosystems uses organic substances and structures (so-called "biobricks"). – This means that an essential difference between living and purely technical systems would no longer be recognizable, in that "life" would coincide with "perfectly orthogonal *and* autopoietic organization". But essential for natural life, as we know it, are autopoiesis, non-triviality and self-determination, which seems to be principally incompatible with strictly modular orthogonality, triviality and heteronomy.

11 Compare the similar considerations of Brenner (2012).

Nevertheless, it is conceivable that synthetic biologists will design biosystems which are, on the one hand, *autopoietic* and, on the other, *trivial*. However, a "trivialized autopoiesis" would originate a new type of autopoiesis (as well as of "machine"), because within the autopoiesis of natural living beings orthogonality only plays a minor part.[12] Moreover, such systems with a "trivialized autopoiesis" would not be able to participate in the game of evolution because they would not be adaptable to a real "struggle for life". In this case, an important basic property (adaptability) which defines *natural* life would be missing. But the question of whether such artificial systems would be "living systems" in the full sense, remains open, as well as the ethical question of which peculiar responsibility we would have towards such engineered and *de novo* designed organisms.

References

Bhutkar, A. (2005): Synthetic Biology: Navigating the Challenge Ahead. In: *J. Biolaw & Bus.*, vol. 8 (2), pp. 19–20.

Billerbeck, Sonja/Panke, Sven (2012): Synthetische Biologie – Biotechnologie als eine Ingenieurwissenschaft. In: Boldt, Joachim/Müller, Oliver/Maio, Giovanni (eds.): *Leben schaffen? Philosophische und ethische Reflexionen zur Synthetischen Biologie*. Paderborn, pp. 19–40.

Boldt, Joachim/Müller, Oliver/Maio, Giovanni (2009), *Synthetische Biologie. Eine ethisch-philosophische Analyse*. Bern.

Brenner, Andreas (2012): Leben leben und Leben machen. Die Synthetische Biologie als Herausforderung für die Frage nach dem Lebensbegriff. In: Boldt, Joachim/Müller, Oliver/Maio, Giovanni (eds.): *Leben schaffen? Philosophische und ethische Reflexionen zur Synthetischen Biologie*. Paderborn, pp. 105–120.

Deplazes, Anna (2008): Nanobiotechnology and Synthetic Biology. In: Ach, Johann S.; Weidemann, Christian (eds.): *Size Matters – Ethical, Legal and Social Aspects of Nanobiotechnology and Nano-Medicine*. Müensteraner Bioethik-Studien, vol. 8, pp. 51–71.

Endy, D. (2005): Foundations for engineering biology. In: *Nature*, vol. 438, pp. 449–453.

European Commission (2005): *Synthetic Biology: Applying Engineering to Biology. Report of a NEST High-Level Expert Group*. Brussels, Belgium.

Frischknecht, F. (2003): The history of biological warfare. In: *EMBO Reports*, vol. 4, pp. 547–552.

Hoffmeyer, Jesper (ed.) (2008): *A Legacy for Living Systems: Gregory Bateson as a Precursor to Biosemiotics*. Berlin.

Jones, R. (2008): Will nanotechnology lead to a truly synthetic biology? See: http://www.softmachines.org/wordpress/?p=440 (last access at 16. Oct. 2011)

Kaiser, J. (2007): Synthetic biology. Attempt to patent artificial organism draws a protest. In: *Science*, vol. 316, p. 1557.

---

12 Natural autopoiesis has a high degree of internal cross-overs (connectivity) and multi-functionality of subsystems (molecular components, cells, tissues, organs etc.).

Kelle, Alexander (2009): Ensuring the security of synthetic biology – towards a 5P governance strategy. In: *Syst. Synth. Biol.*, vol. 3, pp. 85–90.

Kull, Kalevi (1993): *Lectures in Theoretical Biology: The Second Stage*. Tallinn: Estonian Academy of Scineces, pp. 52–62.

Leduc, S. (1912): La biologie synthétique. In: Poinat, A. (ed.): Études de bioph*ysique*, Paris.

Martin, V. J.; Pitera, D. J.; Withers, S. T.; Newman, J. D.; Keasling, J. D. (2003): Engineering a mevalonate pathway in Escherichia coli for production of terpenoids. In: *Nat. Biotechnol.*, vol. 21, pp. 796–802.

Maturana, Humberto/Varela, Francisco (1980): *Autopoiesis and Cognition: The Realization of the Living*. Boston.

Motavalli, J. (2010): "Synthetic Biology" Sparks Nano-Terror Scenario. See: http://newhaveninde-pendent.org/index.php/archives/entry/synthetic_biology_sparks (last access at 15. Oct. 2012).

Rutz, B. (2009): Synthetic biology and patents – A European perspective. In: *EMBO Reports*, vol. 10, special issue, pp. S 14–17.

Schmidt. Markus (2012): Xenobiologie: Neues Leben mit integrierter Biosicherrheit. In: Boldt, Joachim/Müller, Oliver/Maio, Giovanni (eds.): *Leben schaffen? Philosophische und ethische Reflexionen zur Synthetischen Biologie*. Paderborn, pp. 65–79.

Schummer, Joachim (2011): *Das Gotteshandwerk. Die kuenstliche Herstellung von Leben im Labor*. Frankfurt/Main.

Sebeok, Thomas A./Hoffmeyer, Jesper/Emmercehe, Claus (eds.) (1999): *Biosemiotica*. Berlin/ New York [Semiotica vol 127 (no. 1–4)].

Specter, M. (2009): A Life of Its Own. In: *New Yorker*, 28. Sept., pp. 56.

Szybalski, W. (1974): In vivo and in vitro Initiation of transcription. In: *Adv. Exp. Med. Biol.*, vol. 44, pp. 23–24.

The BioBricks Foundation (2010): http://bbf.openwetware.org/ (last access at 10. Oct. 2012).

Then, Christop/Hamberger, Sylvia (2010), *Synthetische Biologie und künstliches Leben – Eine kritische Analyse*. Edited by Testbiotech e.V., München.

Torrance, A. W. (2010): Synthesizing Law for Synthetic Biology. In: *Minnesota Journal of Law, Science and Technology*, 2, pp. 629–665.

TRIPS (1994): *Agreement on Trade related Aspects of Intellectual Property Rigths*. See: http:// www.wto.org/english/docs_e/legal_e/27-trips.pdf (last access at 10. Oct. 2012).

Tucker, J. B.; Zilinskas, R. A. (2006): The promise and perils of synthetic biology. In: *New Atlantis*, 12, pp. 25–45.

von Foerster, Heinz (1999): *Sicht und Einsicht: Versuche zu einer operativen Erkenntnistheorie*. Heidelberg 1999 [1. ed. Braunschweig 1985].

Different Legal Regulations on Life Sciences in Europe and Worldwide

Protecting Humans as Consumers and Subjects: Different Ethical Standards in the Conduct of Research in Europe and America

Jennifer Radden

1. Introduction

The new and emerging science and technology (NEST) of today leaves most people gasping – at the way it will affect every corner of our lives, at its speed in effecting change, and at its transformational potential. This discussion is about some of the ethical underpinnings of policies around protecting humans, as consumers and as subjects, in the face of these breath-taking new realities. My focus is policy and values differences between, and within, Western nations over particular issues. Politics and practicalities, as well as local idiosyncrasies and other non-ethically based cultural variations explain many, but not all, such differences. About those that remain, it can be asked *on what value presuppositions,* and *within what form of ethical reasoning,* are they grounded?

One example will illustrate. It is a feature of certain ethical stances but not others to judge acts of omission as harshly as those of commission when harmful consequences ensue. Consider the rousing imperative of Philip Bond, US Undersecretary of commerce at a SwissRe workshop in Dec 2004, speaking of the potential of nanotechnology:

> "[…] it would be unethical…to attempt to halt scientific and technological progress in nanotechnology. […] Given [its] fantastic potential, how can our attempt to harness nanotechnology's power at the earliest opportunity – to alleviate so many earthly ills – be anything other than ethical? *Conversely, how can a choice to halt be anything other than unethical?*" (emphasis added)

Hyperbolic rhetoric, perhaps. But taken seriously, and *from some ethical perspectives*, this conclusion may be unarguably true, while from others it would not (prudence may dictate moving forward with all possible speed on nanotechnology from those others, but not ethics).

Decisions about protecting humans as consumers and as subjects regularly implicate these deeper-level, moral and ethical variations of presupposition and approach, although such differences are more often alluded to by appeal to slogans and metaphors than formal arguments. (References to playing God, quick technological "fixes," and to Prometheus, for example, stand in for fuller positions.) While not the only ethical approaches possible, four such variations are commonly found in references to NEST issues. (1) Policy may be shaped by the traditional, deontological approach associated with Kantian ethics that is couched in terms of particular, absolute or non-absolute, responsibilities, rights and duties. Instead (2), it may be guided by questions of global justice and an equal distribution of burdens and benefits along the lines of Rawls' difference principle (Rawls 1976). Then again (3), predicted outcomes might be the focus, including uncertainties, and a utilitarian calculation of likely costs, benefits and risks. Finally (4), appeal might be to fundamental conceptions of the person, the good life and nature of human flourishing, as we sometimes find in evaluations of enhancement technologies.[1]

Several points can be made about these alternative approaches. First, whichever one is employed *may result in the same substantive conclusion* about first-order ethical questions. The utilitarian, the Kantian, the Rawlsian might agree on what is the ethical way to proceed in policy terms, even if they cannot agree on the proper way to reason about it. From the perspective of policy decisions, these differences will be of marginal significance. Yet other differences, themselves sometimes irreconcilable, lead towards contrary policy recommendations, and it is these that are my present focus.

Choosing between these approaches also involves a question, although a second-order - and very difficult - one not embarked on here. (Questions such as these involve what is known as meta-ethics.) But, and this is my second point: once we distinguish first-order from second-order questions this way, we encounter other second-order ethical questions more immediately implicated in policy decisions, such as *who decides* policy when it comes to addressing new and emerging science and technology issues, and where the burden of responsibility lies for such policy decisions and their sequelae. (These second- order questions are noted briefly in Part 2.)

A last point is that ethical differences are not always immediately apparent due to the veneer of consensus provided by human rights discourse. Although a rhetorically powerful and useful way of arguing about substantive matters, rights talk sometimes obscures differences reflecting distinctive ethical starting points. Indeed, such discourse is flexible enough to accommodate most different ethical

1 See Sandel 2007.

presuppositions and approaches. Deontological frameworks from the Kantian tradition are the original home of rights talk, but consequentialist and other non-deontological arguments are also sometimes expressed as a demand for equal rights. In some cases, perhaps, these seemingly mixed stances will reflect just that: a blend of deontological and non-deontological concerns that are ethically wanting only in their inconsistency. Nonetheless, the generic flexibility of rights discourse must not blind us to some real differences of presupposition and approach that explain policy variations.

2. Part I

Setting aside the more complex second-order question of which among these competing forms of moral reasoning is the right one to adopt, we can begin with some differences in first- order, substantive, ethical questions. I want to look at three controversies in turn: embryonic stem cell research; genetically modified (GM) food and feed; and some issues that have arisen over experimental research on human subjects.

2.1 Case 1: Embryonic stem cell research

"[...] even the most noble ends do not justify any means [...] the fact that a living being is going to die does not justify experimenting on it or exploiting it as a natural resource." (George Bush, 2001)

The battle lines in this case are evident. The view that embryonic stem cell research is impermissible is conveyed in the above quotation. That such research is permissible is reflected in policies from European nations, and in President Obama's 2009 executive order reversing the Bush prohibition.

The ethical considerations have also been fairly clearly articulated over this set of policies. In favor of stem cell research is the conviction that it holds promise for the treatment and cure of devastating conditions. Against the research, there are two common arguments, that (i) the destruction of human embryos which should be regarded as having the same moral inviolability as fully developed human beings; and (ii) a "slippery slope" will lead to further unacceptable situations (embryo farms, cloning, the use of fetuses for spare parts, and the commodification of human life).

This difference seems to derive from genuinely incompatible ideas about human life in (i), often buttressed by appeal to human rights, especially the right to life understood in absolutist, deontological terms. No compromise position will easily be found. These are fundamental tenets of belief and would need to be addressed directly, if we hope to persuade the person holding them to change her mind on policy governing stem cell research. By contrast, notice that the slippery slope argument in (ii) involves an appeal to outcomes, i.e., it is a form of consequentialism. Dissuading someone from prohibiting stem cell research on the basis of this argument will involve challenging its predictions, or showing ways policy can anticipate and prevent those predicted outcomes. Since arguments in favor of stem cell research also appeal to outcomes and consequences, this second argument "speaks to" them. Disagreements over predicted consequences, including likely outcomes and the weighing of risks, foreseen and unforeseen, are not entirely eliminated by employing consequentialist reasoning, it is true. But they are reduced in one key respect: ruled out, at least, is the likelihood that such disagreements will be a futile exercise employing assumptions so incompatible that they pass each other by.

2.2 Case 2: Genetically modified (GM) food and livestock feed

Differences here are immediately evident, reflected in public awareness and attitudes. In the US, *oblivion* sums up the state of awareness, and an attitude of *trust in government oversight* (primarily, the FDA). By contrast, in Europe, and especially the UK, there is focused awareness and an attitude of *consternation* over GM, as evidenced in scare-mongering terms like "Frankenstein food," and "Frankenfood." These contrary attitudes find expression in policy differences over labeling and regulation. Thus, EU labeling is mandatory for all genetically modified foods if they contain detectable amounts of foreign DNA or protein.[2] By contrast, the US has no mandatory labeling. Moreover, fewer GM foods have been approved at all in Europe than in the US and Canada.

The ethical considerations here have also been loudly proclaimed. GM, it has been said, is a potential answer to world hunger. Yet against that, GM is viewed as a health risk and environmental threat, with unforeseeable long-term consequences. In addition, resistance to GM has sometimes expressed concern over tampering with nature and, in the telling phrase that we often hear in NEST debates, "playing God."

Much of the disagreement here stem froms utilitarian reasoning about the dire and or uncertain consequences of altering food and feed substances this way. (Even within the US, it should be added, consensus is lacking. These attitudes have been shown to vary by gender (women are more concerned); age (the young are less concerned) and familiarity with biotechnology (the better informed are less concerned. (See Evenson 2004)) Admonitions against playing God, however, introduce

---

2 See EC # 1829/2003.

a different set of presuppositions. These appeal to our fundamental nature as human creatures whose limited understanding and moral stature, because they are somehow divinely or naturally ordained, must remain fixed.

The objections to GM are often couched in terms of rights: on the labeling issue, people are said to have *a right to know* about what we ingest and feed livestock; on the matter of regulation, it is said that if genetically modified food and feed are, or are likely to be, harmful, then we have a *right to be protected* from them. But, much of the underlying reasoning is consequentialist nonetheless, whether on the part of opponents of genetically modified foods, who stress the costs and risks, or supporters, emphasizing the benefits and opportunities.

2.3 Case 3: Experimental research on human subjects

Ethical research on human subjects involves what is variously called the issue of clinical equipoise, or the "standard of care" debate. The World Medical Association 2004 Helsinki doctrine is expressed Paragraph 29 of the document widely regarded as containing the most authoritative about research involving human subjects. The central idea of that paragraph is that in any experiment on humans, the experimental intervention and the best-proven therapy should appear to the experimenter to be equally effective (whatever she may hope or wish). Paragraph 29 states: "The benefits, risks, burdens and effectiveness of a new method should be tested against those of the best current prophylactic, diagnostic, and therapeutic methods."[3] So a randomized trial is ethical only when there is genuine uncertainty (or clinical equipoise) as to whether any of the treatments offered is superior. (In keeping with this reasoning studies have been cancelled before their planned completion when the new treatment under consideration showed itself to be vastly more effective.)

This is a particularly far-reaching case. Since all new drugs and medical products must be tested on human subjects before they are sold, experiments on human subjects form a seamless element of the product development chain. Its ethical significance is equally important due to the vulnerability of today's research subjects, fewer and fewer of whom are enlisted in Western countries. (It is more convenient and efficient for companies to conduct studies in developing nations.)[4,5]

---

3 For some history of the evolving wording in WMA documents about the ethics of research on human subjects, see Carlson, Boyd and Webb 2004.

4 In 2009, for example, the Association of Clinical Research Organizations (ACRO) conducted more than 9,000 clinical trials in 115 countries. And the cost effectiveness of this approach is vast: in a 2009 study by the ACRO it was estimated that while a US-based trial would take 5.8 years to enroll, going global would reduce that to 1.9 years.

5 For a critique of these trends, see Garrafa et al. 2010.

Paragraph 29 of the Helsinki Declaration is immediately followed by Paragraph 30, stating that: "At the conclusion of the study, every patient entered into the study should be assured of access to the best proven prophylactic, diagnostic and therapeutic methods identified in the study." On one interpretation of these words, it has been said that: "Research participants in developing countries should receive beneficial new drugs even if this implies supplying the drug for a lifetime if necessary there should be an obligation that the clinical trial benefits the community that contributed to the development of the drug." (European Group in Ethics in Science and New Technologies 2003).[6]

Resistance to the Helsinki declaration has been fierce and wide-ranging.[7,8] One critique attributes a failure to distinguish ethical issues from issues of social justice, for example.[9] Yet for defenders of the declaration, it seems obvious, this will be a distinction without a difference: on some ethical principles such as the Rawlsian one, ethical responses *encompass* and *cannot be detached from* considerations of social justice.

6 For a discussion of the extensive controversy over this issue, see Schüklenk 2004.

7 In the US, the FDA chose to refer to the 1989 version of the Declaration, which the WMA considers invalid, and to adhere to the Good Clinical Practice: Consolidated Guidelines, a joint effort by drug approval agencies from Europe, the USA and Japan, and pharmaceutical trade organizations, which names and honors, but diverges from and weakens, the wording of the Helsinki document (ICH 1996). The European Commission, which sets the pace on national drug laws and ethical standards in Europe, has shown itself similarly reluctant to accept Helsinki, and instead referred to the 1996 version in its 2001 clinical trials directive (EC 2001) and its 2005 directive on good clinical practices (EC 2005).

8 In light of such critique, changes to the Declaration have been made. Footnotes now added to 29 and 30 of the Declaration of Helsinki in light of these criticisms include: "a placebo controlled trial may be ethically acceptable, even if proven therapy is available under some circumstances" and "it is necessary during the study planning process to identify post-trial access by study participants to prophylactic, diagnostic and therapeutic procedures identified as beneficial in the study or access to other appropriate care. Post-trial access arrangements or other care must be described in the study protocol so the ethical review committee may consider such arrangements during its review" (WMA 2004).

9 "… the Declaration has moved from a purely ethical document to a document that is increasingly interested in social justice…The WMA is clearly very upset that people in poor countries don't have really good medical care. And I am upset by that too. But I don't think that determines the ethics of the trial" (Wolinsky 2006: 670).

This debate often comprises a rather inflammable confusion of ideas.[10] The question of what is owed experimental subjects after studies have been completed is clearly an ethical one. But it can be separated from whether equipoise should be seen as ethically required of any protocol. What appears to blur this line in many minds is that the Helsinki position with regard to each rests on underlying moral reasoning that emphasizes equality, and an allocation of burdens and benefits that is above all fairly distributed, i.e., a Rawlsian conception of global justice.

Even disengaged from the issues raised by Paragraph 30, the case for clinical equipoise is difficult to parse, with its mixture of ethical science and science ethics. At first glance, Paragraph 29 appears to be a methodological or scientific, rather than an ethical stance – except shoddy science, arguably, is itself unethical. The clinical equipoise requiring that investigators not know which of two treatments subjected to clinical trial is more effective is called for because otherwise they would be guilty not only of knowingly giving inferior treatment to some of the participants, but of engaging in scientifically redundant, and thus ethically unwarranted, exercises (Angell 1997).

Some critics have insisted that basic questions of validity are at risk in experiments that eliminate or limit the placebo arm, as Paragraph 29 originally did. (Its qualifying footnote in the text removes this blanket prohibition. See footnote 8, page XXX.) This becomes important in evaluating disorders known to be responsive to placebo effects, including conditions such as depression. The point has recently been emphasized by Robert Temple, director of the office of medical policy at the FDA's Center for Drug Evaluation and Research: "[…] you really *do need placebos* to learn something…in a lot of symptomatic conditions […] " (Temple 2010).[11] *Only* by including a placebo arm when assessing the effectiveness of many classes of drugs, this suggests, will we discover their effectiveness. In light of these and other criticisms, the principle of clinical equipoise enunciated

10 The principle of clinical equipoise has engendered such confusion and dissent that one proposal eliminates it altogether, instead proposing that all research on human subjects adhere to seven particular requirements. To be ethical research must have: (i) scientific or social value; (ii) scientific validity; (iii) fair subject selection; (iv) favorable risk/benefit ratio; (v) independent review; (vi) informed consent, and finally, (vii) respect for the enrolled research participants. See Emmanuel 2008.

11 "You can sometimes use a NI study design where there is a solid basis for knowing the effect of the positive control, but that would be impossible in depression, anxiety, and most symptomatic conditions; for those you need a placebo to show assay sensitivity, i.e., that you can tell one thing from another, because many studies in those conditions cannot tell active drugs from placebo." (www.fda.gov/downloads/Drugs/NewsEvents/UCM209270)

in Paragraph 29 has been judged best applied selectively, depending on the particular research protocol.[12]

A critique that is directed at claims in Paragraph 30 focuses on the quality of voluntary consent (ethically) required from human subjects. In a discussion paper on the revised Declaration of Helsinki from the Pharmaceutical Research and Manufacturers of America it has been pointed out that promising care to study participants would amount to "undue" – for which we can read unjust or unethical – inducement (PhRMA 2001). Inasmuch as this would jeopardize the voluntariness of the participants' agreement to take part, this is a serious consideration, whatever its motivation. (Because PhRMA is a trade association representing the country's research-based drug industry, its impartiality cannot be accepted uncritically.)

Employing related ethical presuppositions, and again focused on the issue of the participants' consent, some have emphasized the special role of the research subject. Good research is by its nature somewhat exploitative of individual research subjects, it is insisted, making them different from patients in therapeutic endeavors. Their informed consent protects such subjects as far as possible, but of necessity cannot entirely insulate them from risk. In demanding clinical equipoise, the Declaration of Helsinki has been accused of losing sight of this fact and of the moral status of research participants as informed, voluntary agents. (See Wendler 2002.)

A final example of incompatible ethical assumptions comes from debates over Paragraph 30, with its underlying appeal to a sense of justice that emphasizes equality, and an allocation of burdens and benefits that is above all fairly distributed. Critiques, now acknowledged and somewhat answered in a qualifying footnote added to the 2004 version (see footnote 8), have cautioned against treating the better as the enemy of best:

> "Surely it is better for a seriously ill person to receive a good drug, even if it is not the best, than to receive no drug at all." (Brock and Winkler 2006)

The gradualist position expressed here - that *any improvement is better than none* - is contradicted by, and incompatible with, the absolutism embodied in calls for an equal distribution of burdens and benefits.

Of the three cases selected for discussion each exhibits first rank ethical disagreement. But – as long as the opponents are working within the same conception of how to proceed, these are in certain respects the least difficult to deal with. The view that stem cell research is justified because it will improve the lives of more people with terrible diseases than there are fetuses sacrificed, or that fetal life is

12 See Carpenter, Appelbaum and Levine 2003.

not, or not as, valuable, may not ever be compatible with the one that there is an inviolable fetal right to life. By a utilitarian calculation, the future medical benefits for many people *outweigh* the insignificant sacrifice of some fetal lives. Similarly, in the human subjects example, when it challenges whether proper informed consent is compatible with entirely protecting subjects from exploitation, or accuses the Helsinki Declaration of treating the better as the enemy of best, opposition to Paragraph 30 introduces incompatible ethical principles.

So even these first order, substantive disagreements are complicated by the fact that different traditions of ethical reasoning embody different assumptions and ways to proceed. Whichever approach is employed (appeal to responsibilities and duties, to global justice and equality, to consequences in terms of likely costs and benefits, or to fundamental ideas and conceptions about human nature), it may result in the same substantive conclusion. But it may not.

3. Part II

Second-order ethical questions include the very hoary one of which ethical approach to adopt which we cannot hope to resolve it here. Yet other second order questions that might affect policy differences are not so easily set aside: two of these, which are briefly introduced below, ask *who decides*, and *where responsibilities lie*, when policies over protecting human consumers and subjects are at issue.

Economies of scale, and political and economic structures, dictate that governments rather than individuals put policies in place, but who determines what those policies are to be? When it comes to NEST issues several competing models suggest themselves, the first focused on the value of expert knowledge. Panels of experts, convened by governments, seem the best positioned to deal with the technical complexity of NEST issues. Arguably, only such experts have the knowledge to do so, and certainly they have a knowledge *advantage* over non-experts (See Halal 2008, Weinberger 2011.) Yet, if scientific facts are understood to be inaccessible to non-experts, ethical considerations are not. And in the face of public concerns, it is seemingly the duty of the policy implementers to listen; after all, these NEST innovations will affect the individuals making up that public quite as much as it will the policy implementers. The idea of deliberative democracy (Habermas) is embodied in "consensus conferences" or "participatory technology assessment," where a small group of lay citizens examine a controversial technological issue to the best of their capability, identifying key areas, cross-examining experts, and then arriving at a consensus position presented to policy makers and to the general public.[13,14] These have been used for more than a decade in Denmark, and have since been tried in Australia and Canada in addressing issues of food biotechnology.[15] Increasingly their use has been advocated in a range of efforts to engage citizens in the questions of science and technology that will affect their lives so profoundly. One of the most obvious and remarked limitations of consensus conferences pertains to how well they represent of the population from which they are drawn, but other issues arise as well.[16] Expert models are pitted against consensus models over the question of whether ethical knowledge is to be privileged over superior scientific knowhow.

The question of who should make policy about the ethical implications of NEST innovations then, is complex and far from resolved. And where the burden of responsibility lies is an allied issue with its own fraught ethical challenges, including that of self-regulation. The call for a voluntary moratorium on recombinant DNA research in 1974 and 1975 by molecular biologists illustrates a model favoring self-regulation. So do Institutional Review Boards (otherwise known as IECs (Independent Ethics Committees), ERBs (Ethical Review Boards) or Committees for the protection of human subjects)), that give scientists and research institutions regulative responsibility in relation to research. Requiring elements of infrastructure and expertise not likely present in the developing world settings where research trials are conducted, or in smaller institutions in the West,[17] such boards also have a number of obvious drawbacks. In addition, they remain prey to conflicts of interest inasmuch as they are placed *within* the research institutions (Sugarman 2000). Both forms of self-regulation, and of external regulation seem to have inherent vulnerabilities.[18]

These considerations are beyond the scope of the present discussion. But just as we saw in relation to first-order, substantive disagreements, it seems likely that answers to these second-order questions about who should make policy, and where responsibilities lie in relation to it, will sometimes also diverge along lines that reflect deep, ethical differences of approach and presupposition.

---

13 For Habermas's notion of deliberative democracy, see Habermas, Cronin and De Greiff 2000, Benhabib 1996.

14 See Joss and Durant 1995.

15 See Powell and Kleinman 2008, Einsiedel, Jelson, and Breck, 2001.

16 See Kleinman, Powell, Grice et. al. 2007, Powell and Colin 2008.

17 It has been pointed out that:[1] "poor and low-income countries, which lack sustainable control and review systems to deal with the ethical and methodological challenges of complex studies conducted by researchers from affluent countries and funded by large multinational pharmaceutical companies, are particularly vulnerable." (Garrafa et al. 2010)

18 See Phillips 1996.

4. Conclusion

Summing up then, I want to emphasize several points. First, policy differences sometimes emerge from incompatibilities between differing ethical frameworks and presuppositions, and these incompatibilities need to be recognized to avoid debates that are fruitless. This is not to deny that less ethical considerations account for these policy differences as well, or that differences do not often remain even between those sharing the same ethical position. Second, as long as we recognize the need for consistency within our reasoning, we must conclude that while rhetorically valuable, references to human rights sometimes serve to conceal fundamental incompatibilities among ethical ideas. And finally, as well as first order ethical questions about biotechnology and humans as subjects and consumers, there are important second-order questions to be addressed as we try to make sense of this new and emerging science and technology. Answers must be found over who decides policy and where the burden of responsibility lies for doing the right thing – which must include who or what is to blame when disasters occur. And these answers too, I would suggest, will sometimes be influenced by fundamental ethical differences.

Acknowledgements

This paper was originally presented as 'Different Ethical Standards in the Conduct of Research in Europe versus America' at the Life Sciences University Network Workshop, Berlin February 2010. For their support I am grateful to Prof. Dr. Hilmar Stolte and Dr. Eunsook Hyun. My thanks also go to participants at that workshop for their illuminating and helpful comments.

References

Angell, M. (1997): The Ethicsl of Clinical Research in the Third World. In: *N Engl J Med* 337: pp. 847–849.

Benhabib, S. (ed.) (1996): *Democracy and Difference: Contesting the Boundaries of the Political.* Princeton, NJ: Princeton University Press.

Brock, D. and Winkler, D. (2006): Ethical Issues in Resource Allocation, Research and New Product Development, Chapter 14, in Jamison, Dean, et.al. (eds.), *Disease Control Priorities in Developing Countries*. World Bank. Disease Control Priorities project, pp. 259–270.

Carlson, R., Boyd, K. and Webb, D. (2004): The revision of the Declaration of Helsinki: past, present and future. In: *Br. J. Clin. Pharmacol.* 57: pp. 695–713.

Carpenter, W., Appelbaum, P. and Levine, R. (2003): The Declaration of Helsinki and Clinical Trials: A Focus on Placebo-Controlled Trials in Schizophrenia. In: *Am.J. Psychiatry* 160: pp. 356–362.

Einsiedel, E., Jelson, E. and Breck, T. (2001): Publics at the technology table: The consensus conference in Denmark, Canada, and Australia. In: *Public Understanding of Science* 10: pp. 83–98.

Emmanuel, E. (2008): Clinical Research Training, National Institutes of Health. Washington DC: NIH.

Evenson, R.E. (ed.) (2004): *Consumer Acceptance of Genetically Modified Foods*. CABI Publishing.

Garrafa, V., Solbakk, J. H., Vidal, S. and Lorenzo, C. (2010): Between the needy and the greedy: the quest for a just and fair ethics of clinical research. In: *J. Med. Ethics* 2010; 36:8, pp. 500–504.

Habermas, J., Cronin, C. and De Greiff, P. (eds.) (2000): *The Inclusion of the Other: Studies in Political Theory*. MIT Press.

Halal, W. (2008): *Technology's Promise: Expert Knowledge on the Transformation of Business and Society.* Palgrave, Macmillan.

Joss, S. and Durant, J. (1995): *Public Participation in Science: the Role of Consensus Conferences in Europe.* NMSI Trading Ltd.

Kleinman, D., Powell, M., Grice, J. et al. (2007): A Toolkit for Democratizing Science and Technology Policy: The Practical Mechanics of Organizing a Consensus Conference. In: *Bulletin of Science, Technology, Society*. 27: pp. 154–69.

Phillips, D. F. (1996): Institutional Review Boards Under Stress: Will They Explode or Change? In: *JAMA* 276: pp. 1623–26.

Powell, M. and Kleinman, D. L. (2008): Building citizen capacities for participation in nanotechnology decision-making: the democratic virtues of the consensus Conference model. In: *Public Understanding of Science* 17: pp. 329–348.

Powell, M. and Colin, M. (2008): Meaningful Citizen Engagement in Science and Technology. What Would It Really Take? In: *Science Communication*. 30: pp. 126–36.

Rawls, J. (1976): *A Theory of Justice*. Harvard: Harvard University Press.

Sandel, M. S. (2007): *The Case Against Perfection: Ethics in the Age of Genetic Engineering*. Harvard: Harvard University Press.

Schüklenk, U. (2004): The standard of care debate: against the myth of an "international consensus opinion." *Journal of Medical Ethics* 30: pp. 194–97.

Sugarman, J. (2000): The Role of Institutional Support in Protecting Human Research Subjects. In: *Academic Medicine*. 75: pp. 687–92.

Weinberger, D. (2011): *Too Big to Know: Rethinking Knowledge Now That the Facts Aren't the Facts, Experts Are Everywhere, and the Smartest Person in the Room Is the Room*. New York: Basic Books.

Wendler, D. (2002): Informed Consent, Exploitation, and Whether it is Possible to Conduct Human Subjects Research Without Either One. In: *Bioethics* 14: pp. 310–339.

Wolinsky, H. (2006): The Battle of Helsinki. In: *EMBO Reports* 7: pp. 670–72.

Legal Regulation of Genetic Information in the USA and Germany – Promises kept in Insurance?

*Julian Ziegler**

1. Introduction

Human rights, human dignity and individual privacy rights are terms which all regularly surface in the international discussion of ethical and legal consequences of molecular biology and which are also frequently used in the international discussion of genetic information, as well as its usage in private insurance underwriting. Especially in the United States of America (USA), where the vast majority of the population consists of privately insured individuals, either through group insurance plans as fringe benefits of employment or through individually purchased insurance plans, the question of whether insurers should be allowed access to and use of genetic information of individual insurance applicants or those already insured, has been under discussion for years. Finally, after more than twelve years of legislative proceedings, a piece of legislation dealing specifically with genetic information, the Genetic Information Nondiscrimination Act of 2008 (GINA)[1], was signed into law by President George W. Bush on 21st May 2008. In Germany, however, the political discussion about whether specific legislation was needed to regulate the use of genetic information took until 21st February 2010, the date the German Gene Diagnostics Act[2] was finally enacted.

Both laws, GINA and the GenDG, were fueled by public fears of discrimination based on individuals' genes and consequently address these fears, mainly by prohibiting access to and use of genetic information by private insurance providers. However, the way they regulate the use of genetic information and its use by insurance companies and the scope of protection between those two laws differ in detail.

---

* Rechtsassessor Dr. iur. Julian Ziegler, Hamburg.

1 Genetic Information Nondiscrimination Act of 2008, Pub. L., No. 110–233, 122 Stat. 881 (2008).

2 Gesetz über genetische Untersuchungen beim Menschen (Gendiagnostikgesetz – GenDG), vom 31.07.2009 (BGBl. 2009 I, 2529).

The purpose of this contribution is to point out these regulatory differences, as well as to examine whether protection against genetic discrimination in the private insurance context meets the expectations connected with the enactment of the respective laws.

2. Insurers' interest in genetic information

Especially in the context of private insurance contracts, public fears of discrimination based on individuals' genes become virulent. As genetic information has the potential to improve the predictability of health risks[3], its use is potentially valuable for insurers who aim at an actuarially fair[4] calculation of insurance plans divided according to individual risk.[5] For this reason, during the underwriting process a risk-classification of insurance applicants according to sex, age and individual health status is undertaken in order to establish insurance plans (or tariffs) with individuals representing similar health risks.[6] As this only works in a situation where an insurance applicant and the insurer share the same knowledge about the individual health risks, insurers demand disclosure of information from the applicants about their health status prior to concluding the insurance contract. Should the insurer know less about individual risk factors than the respective insured,

3 Kerstin Berberich, Zur Zulässigkeit genetischer Tests in der Lebens- und privaten Krankenversicherung, Albrecht, P, Lorenz, E., (1998), 284; Jochen Taupitz, Genetische Diagnostik und Versicherungsrecht, Verlag Versicherungswirtschaft, (2000), 31–32; Julian Ziegler, Genanalysen in der Deutschen Privaten Krankenversicherung, Lang, (2008), 56–78; Oliver Brand, *Grenzen der vorvertraglichen Anzeigepflichten des Versicherungsnehmers*, VersR, (2009), 715–721, 718; Julian Ziegler, *Genetische Diskriminierung in der privaten Krankenversicherung in den USA und Deutschland: Alter Wein in neuen Schläuchen?*, 34 DAJV-Newsletter, (2009), 114–122, 118.

4 Jürgen Prölss, *Der Versicherer als „Treuhänder der Gefahrengemeinschaft" - Zur Wahrnehmung kollektiver Belange der Versicherungsnehmer durch den Privatversicherer, in* Festschrift für Karl Larenz zum 80. Geburtstag, (1983), 487–535, 523–524; Taupitz, 12.

5 Taupitz, 9; Ziegler, 118.

6 Erich R. Prölss & Anton Martin, Versicherungsvertragsgesetz Helmut Kollhosser Ulrich Knappmann, Jürgen Prölss, Wolfgang Voit, Christian Armbrüster ed., C. H. Beck 27 Aufl. ed., (2004), § 178 g, ref. 3–5; Peter Albrecht, *Gesetze der großen Zahlen und Ausgleich im Kollektiv - Bemerkungen zu Grundlagen der Versicherungsproduktion*, ZVersWiss, (1982), 501–538, 511; Hanns Bauermeister & Klaus Bohn, *Krankenversicherungsmathematik, in* Handwörterbuch der Versicherung (HdV), Dieter Farny, et al. eds., (1988), 401–407, 401; Ziegler, 100.

'adverse selection'[7] could occur, where the insured would be able to obtain insurance at a premium lower than would be equivalent to the real individual risk, thus forcing low-risk individuals to subsidize high-risk individuals within the same insurance plan.[8] This in turn could endanger the insurance plan itself.[9]

In the course of the underwriting process, insurers usually ask applicants to disclose already existing and past medical conditions, their medical family history, their sex as well as their age. Applicants, if accepted by the insurer, are thereafter distributed among groups of individuals of the same sex and similar age, as well as with a similar individual health risk factor calculated on the basis of the information obtained during the underwriting process. This discrimination process, if applied to genetic health risk factors of insurance applicants, is known as "genetic discrimination".

3. Legal approach towards genetic discrimination in the USA: GINA

In the USA, GINA was enacted as a means to provide comprehensive protection from genetic discrimination[10] in health insurance and employment. This had thus far not been offered by already existing federal laws.[11] Of these laws, the Americans with Disabilities Act of 1990 (ADA)[12] and the Executive Order 13,145 did not address the market for insurance directly and are only applicable

---

7 Michael Spence & Richard Zeckhauser, *Insurance, information, and Individual Action*, 61 The American Economic Review, (1971), 380–387, 384; Mark J. Browne, *Evidence of adverse selection in the Individual Health Insurance Market*, 59 The Journal of Risk and Insurance, (1992), 13–33, 32; Mark J. Browne & Helene I. Doerpinghaus, *Information Asymmetries and Adverse Selection in the Market for Individual Medical Expense Insurance*, 60 The Journal of Risk and Insurance, (1993), 300–312.

8 Browne, 32; Taupitz, 12–13; Ziegler, 72–76, with additional references.

9 Seminal: George A. Akerlof, *The Market for "Lemons": Quality Uncertainty and the Market Mechanism*, 84 Quarterly Journal of Economics, (1970), 488–500, 489.

10 R. F. Rich & J. Ziegler, *Genetic discrimination in health insurance–comprehensive legal solutions for a (not so) special problem?*, 2 Indiana Health Law Rev, (2005), 1–47, 9.

11 Genetic Information Nondiscrimination Act of 2008, Pub. L. No. 110–233, 122 Stat. 881, (2008), Sec. 2 (5).

12 Americans with Disabilities Act, Title I, 42 U.S.C. § 12101–12213 (2006); for an in-depth description of the legislative history see e.g.: Jessica L. Roberts, *Preempting Discrimination: Lessons from the Genetic Information Nondiscrimination Act*, 63 Vanderbilt Law Review, (2010), 439–490.

to health insurance as a fringe benefit of employment.[13] Specifically for health insurance, prior to GINA, only the Health Insurance Portability and Accountability Act of 1996 (HIPAA)[14] offered protection against discrimination on the basis of genetic information. However, the protection against discrimination offered by HIPAA was not comprehensive. Especially in the market for individual health insurance, insurers were not prohibited from premium calculation on the basis of genetic information and in the determination of the enrollment of 'non-eligible'[15] individuals.[16] But also in the group market, the premiums for a group of insured could still be adjusted on the basis of genetic information of an individual inside the group.

These flaws were addressed by GINA as the law now prohibits insurers from collecting[17] and using genetic information of applicants or enrollees of both group[18] and individual insurance[19] plans for underwriting purposes.[20] In both markets, GINA also prohibits insurers from requesting or requiring insurance

13 See in detail: Rich & Ziegler, 26–27.

14 Health Insurance Portability and Accountability Act of 1996, 29 U.S.C. §§ 1181–82 (2006), 42 U.S.C. §§ 300gg-41 (2006).

15 For a definition see e.g. Rich & Ziegler, 34 at Footnote 157.

16 Id. at 38–39.

17 Jessica L. Roberts, *The Genetic Information Nondiscrimination Act as an Antidiscrimination Law*, 86 Notre Dame Law Review, (2011), 597–648, 622, who stresses that the mere attempt to obtain genetic information by insurers already violates the statute.

18 Genetic Information Nondiscrimination Act of 2008, Pub. L. No. 110–233, 122 Stat. 881, (2008), § 101 (b), 102 (a) (2), (4):
"(1) IN GENERAL A group health plan, and a health insurance issuer offering health insurance coverage in connection with a group health plan, shall not request, require, or purchase genetic information for underwriting purposes (as defined in section 733)."

19 Genetic Information Nondiscrimination Act of 2008, Pub. L. No. 110–233, 122 Stat. 881, (2008), § 102 (b) (1) (B); "(1) IN GENERAL A health insurance issuer offering health insurance coverage in the individual market shall not request, require, or purchase genetic information for underwriting purposes (as defined in section 2791).

20 Genetic Information Nondiscrimination Act of 2008, Pub. L. No. 110–233, 122 Stat. 881, (2008), § 101 (d): "UNDERWRITING PURPOSES The term 'underwriting purposes' means, with respect to any group health plan, or health insurance coverage offered in connection with a group health plan (A) rules for, or determination of, eligibility (including enrollment and continued eligibility) for benefits under the plan or coverage; (B) the computation of premium or contribution amounts under the plan or coverage; (C) the application of any pre-existing condition exclusion under the plan or coverage; and (D) other activities related to the creation, renewal, or replacement of a contract of health insurance or health benefits."

applicants and insured individuals or their respective family members to undergo a genetic test.[21]

The basis for this protection is the broad definition of the key terms[22] 'genetic information' and 'genetic test' employed by GINA.

Under GINA, 'genetic information' is defined as the information derived from an individual's genetic test, the genetic test of a family member of the individual, or the individual's family history, but it does not encompass information about the sex or age of the individual.[23] In this context 'genetic test' means the detection of genotypes, mutations, or chromosomal changes by analysis of DNA, RNA, chromosomes, proteins and metabolites unless it does not detect genotypes, mutations or chromosomal changes, or is directly related to a manifested disease, disorder, or pathological condition that could reasonably be detected by a health care professional with appropriate training and expertise in the field of medicine involved.[24] – The definitions set forth by GINA cover all four approaches[25] of gene analysis:

- Cytogenetic analysis,
- Protein-chemical analysis,
- DNA-analysis, and
- Phenotype analysis,

as long as such analysis is not directly related to an already manifested disease, disorder, or pathological condition of an individual or a family member of such

21 Genetic Information Nondiscrimination Act of 2008, Pub. L. No. 110–233, 122 Stat. 881, (2008), § 101 (b); 102 (a) (2); 102 (b) (1) (B); 103 (b); 104 (b) (1).

22 Roberts, 621.

23 Genetic Information Nondiscrimination Act of 2008, Pub. L. No. 110–233, 122 Stat. 881, (2008), § 101 (d), § 102 (a) (4), § 103 (d), § 104 (b) (1); This definition incorporates "with respect to any individual, any request for, or receipt of, genetic services, or participation in clinical research which includes genetic services, by such individual or any family member of such individual".

24 Genetic Information Nondiscrimination Act of 2008, Pub. L. No. 110–233, 122 Stat. 881, (2008), § 101 (d), § 102 (a) (4), § 103 (d), § 104 (b) (1).

25 Deutsche Forschungsgemeinschaft, *DFG-Senatskommission - Prädiktive genetische Diagnostik - Wissenschaftliche Grundlagen, praktische Umsetzung und soziale Implementierung* (2003), at http://www.dfg.de/aktuelles_presse/reden_stellungnahmen/2003/download/predictive_genetic_diagnosis.pdf, 18; Rainer Paslack & Hilmar Stolte, Gendiagnostik: Stand der Forschung und neue Optionen für Medizin und Wirtschaft, Ministerium für Wirtschaft und Arbeit des Landes Nordrhein-Westfalen, (2003), 25; Ziegler, 43.

individual. This means that, since the enactment of GINA, insurance underwriting under GINA is effectively limited to information about pre-existing medical conditions (whether gene-related or not) as well as information about the sex and age of insurance applicants and enrollees.

In the context of GINA as an anti-discrimination law, this definition framework facilitates a comprehensive ban on the use of any genetic information which might have predictive value for future medical conditions, regardless of its origin, as a means to discriminate (or differentiate) between insurance applicants and enrollees in the insurance underwriting process, which, prior to GINA, was still permissible to some extent both in the group and individual market for health insurance.

4. Legal approach towards genetic discrimination in Germany: GenDG

In Germany, prior to the enactment of the GenDG, no specific law regulating the acquisition and use of genetic information in health insurance existed. Although there was extensive legal debate about whether the use of genetic information for purposes of differentiation between insurance applicants given the pre-GenDG legal framework was lawful[26] or not[27], the German legislator decided to enact a law specifically prohibiting discrimination on the basis of genetic information, as well as safeguarding the right of individual self-determination[28], primarily driven by the desire to prevent the potential danger of genetic discrimination inherent in the use of genetic informatio.[29] The legislator saw information coming from

26 Egon Lorenz, *Zur Berücksichtigung genetischer Tests und ihrer Ergebnisse beim Abschluß von Personenversicherungsverträgen - ein Diskussionsbeitrag, in* Genetische Diagnostik und Versicherungsschutz. Die Situation in Deutschland, Felix Thiele ed., (2000), 14–35, 33; Taupitz, 45; Ziegler, 137, 160.

27 Erwin Deutsch, *Die Genomanalyse: Neue Rechtsprobleme*, 19 ZRP, (1986), 1–4, 3; Erwin Deutsch, *Das Persönlichkeitsrecht des Patienten*, 192 AcP, (1992), 161–180, 169; Jürgen Simon, *Genomanalyse - Anwendungsmöglichkeiten und rechtlicher Regelungsbedarf*, MDR, (1991), 5–14, 14; Peter Präve, *Das Recht des Versicherungsnehmers auf gen-informationelle Selbstbestimmung*, VersR, (1992), 279–284, 282; Oliver Schöffski, *Genetik und Versicherung aus ökonomischer Sicht*, ZVersWiss, (2002), 227–253, 233; Reinhard Damm, *Persönlichkeitsschutz und medizintechnische Entwicklung*, JZ, (1998), 926–938, 932; Reinhard Damm, *Prädiktive Medizin und Patientenautonomie*, MedR, (1999), 437–448, 438.

28 See the purpose of the law in § 1 GenDG.

29 Bundesregierung, Unterrichtung durch die Bundesregierung, Entwurf eines Gesetzes über genetische Untersuchungen bei Menschen (Gendiagnostikgesetz – GenDG) – Drucksache

genetic examinations conducted in a laboratory as being something special, carrying high predictive potential and therefore bearing a risk for social, ethical and eugenic discrimination.[30] All of these factors appeared to mandate special legal attention, as they differentiate information derived from genetic examinations from information which stems from regular medical examinations. However, as the problems associated with the informational quality of genetic examinations are not limited to insurance, the GenDG tries to encompass all fields, in which an infringement of the right of individual self-determination due to genetic examinations might occur, namely: "genetic examinations for medical purposes" (in chapter 2), "genetic examinations for the determination of ancestry" (in chapter 3), "genetic examinations in insurance" (in chapter 4) and "genetic examinations in employment" (in chapter 5).

a. Key legal definitions pursuant to § 3 GenDG

As the German legislator assumes that the major risk potential for the individual right of self-determination is inherent in the technological differences between genetic examinations and regular medical examinations, § 3 GenDG contains the legal definitions of the key terms 'genetic examination' and 'gene analysis' in order to pinpoint what sets them apart from regular medical examinations. Thus, these definitions serve as the basis for the determination of the scope of protection offered by the law.

§ 3 No. 1 GenDG defines the key term 'genetic examination' as a "gene analysis for the determination of genetic traits and/or a prenatal risk determination". According to § 3 No. 2 GenDG, 'gene analysis' is a "cytogenetic analysis, a molecular genetic analysis, as well as a gene-product analysis." Phenotype analyses, however, are not included in the definition and thus are not regarded as gene analyses under the law, even if they reveal genetic information.[31] The reason for this is the legislators' assumption that in particular gene analyses conducted within a laboratory carry the potential for an abuse of genetic information derived from smallest samples which can be gained without the individual's knowledge.[32]

Although § 3 GenDG contains further definitions, especially the differentiation between 'predictive' (§ 3 No. 8 GenDG) and 'diagnostic' (§ 3 No. 7 GenDG)

16/10532 – Gegenäußerung der Bundesregierung zu der Stellungnahme des Bundesrates, Deutscher Bundestag (16. Wahlperiode), BT-Drs. 16/10532 vom 15. 10. 2008, (2008), 1.

30 Id. at 16.

31 Julian Ziegler & Andreas Ziegler, *Gendiagnostikgesetz und Versicherung: Anspruch und Wirklichkeit*, 100 ZVersWiss, (2011), 29–53, 35; Bundesregierung, 17.

32 Bundesregierung, 17.

genetic examinations, these definitions are not relevant in the context of the health insurance provision of the law, as it covers any type of genetic examination or analysis as defined by § 3 Nos. 1 and 2 GenDG.[33]

b. Provision for insurance, § 18 GenDG

The legislative intention to offer a comprehensive protection against any kind of misuse of genetic examinations or analyses becomes apparent in the formulation of § 18 GenDG, the insurance provision contained in chapter 4 of the GenDG. This provision is applicable for all types of insurance 'contracts', although it is mainly aimed at life and health insurance.[34] This means that any non-private (i.e. 'social') insurance plans are exempted from the protection of the provision as they are not based on an insurance contract between the insured and the insurer.[35]

§ 18 I 1 no. 1 GenDG prohibits insurers from demanding that an individual undergo a genetic examination or gene analysis before or after the closure of the insurance contract. Pursuant to § 18 I 1 no. 2 GenDG, insurers are also prohibited from demanding, receiving or utilizing the results or data stemming from such genetic examinations or gene analyses. The second provision, pursuant to § 18 I 2 GenDG, is not applicable for life-, disability-, occupational disability- and long-term care pension insurance contracts with insurance benefits amounting to at least EUR 300,000 in case of a one-time benefit or EUR 30,000 in case of an annual pension.

Notwithstanding these provisions, according to § 18 II GenDG, pre-existing conditions and existing medical conditions of an individual have to be disclosed to insurers by this individual, as in this context the provisions of the German

33 ZIEGLER & ZIEGLER, 39–41, 45; SEBASTIAN KRÖGER, *Das neue Gendiganostikgesetz und seine Auswirkungen auf den rechtlichen Rahmen beim Abschluss von Versicherungsverträgen*, MedR, (2010), 751–756, 752; Differing: CHRISTIAN ARMBRÜSTER, Stellungnahme zum Entwurf eines Gesetzes über genetische Untersuchungen bei Menschen (Gendiagnostikgesetz - GenDG) im Rahmen der Öffentlichen Anhörung des Ausschusses für Gesundheit des Deutschen Bundestages vom 21. Januar 2009, Ausschussdrucksache 16(14)0469(41) Ausschuss für Gesundheit ed., Deutscher Bundestag, (2009), 7–8.

34 BUNDESRAT, Gesetzentwurf der Bundesregierung - Entwurf eines Gesetzes über genetische Untersuchungen bei Menschen (Gendiagnostikgesetz - GenDG), Drucksache 633/08 vom 29.08.2008, (2008), 73.

35 BUNDESREGIERUNG, 36; KAI-JOCHEN NEUHAUS, *Aktuelle Probleme in der Personenversicherung - unter besonderer Berücksichtigung der Berufsunfähigkeitsversicherung -*, r + s, (2009), 309–319, 315.

Insurance Contract Act (Versicherungsvertragsgesetz – VVG) regarding disclosure obligations of insurance applicants (§§ 19–22 and 47 VVG) still apply.

On first glance, the protections from genetic discrimination contained in § 18 GenDG seem comprehensive enough, as both the demand for genetic examination or analyses and the use of genetic information are banned in insurance. However, severe flaws of the law become apparent once the provision is analyzed in detail.

c. § 18 I 1 GenDG

The provisions contained in § 18 I 1 GenDG are intended as a protection of insurance applicants and insured individuals against the use of their genetic information mainly within the underwriting process. Therefore, individuals may neither be required by insurers to undergo genetic examinations or gene analyses before or after closure of an insurance contract, nor may insurers demand, receive[36] or use information or any kind of data derived in connection with such genetic examinations or analyses.

As the law is directed against discrimination in the underwriting process, the first questionable item of the legal provision is the inclusion of the time *after* closure of the insurance contract.

§ 18 I 1 GenDG includes the time period after closure of an insurance contract under the scope of its protection as a reaction to the possibility in many types of insurance to increase premiums on the basis of health associated risk which become apparent during the life cycle of an insurance contract. This possibility to react to perceived risk increases, e.g. in life insurance contracts[37], however, is not provided for in the German private health insurance system.

Under German law, a private health insurance contract may not be prematurely terminated by the insurer due to changes in the health status of the insured.[38] Correspondingly, changes in the health status of insured may not lead to premium increases for the respective individual. Premium increases are only possible for a whole insurance plan on the basis of changes of the mortality tables of the group as

---

36 For a discussion of the term "receiving" of information or data from genetic examinations or data see: ZIEGLER & ZIEGLER, 48; KRÖGER, 753, according to whom the term requires only a possibility of receiving knowledge of the protected information; a differing opinion is expressed by NEUHAUS, 316; PETER PRÄVE, *Das Gendiagnostikgesetz aus versicherungsrechtlicher Sicht*, 60 VersR, (2009), 857–912, 860, who argue that "receiving" under the law requires an intent of the insurer to use the information.

37 § 163 I VVG.

38 § 206 I 1 VVG.

a whole.[39] Other than that, there is no possibility for changes of the premium rates in private health insurance in Germany.

The other reason for the inclusion of the time after closure of the insurance contract is the legislative interest to prohibit any possibility of those already insured undergoing genetic examinations or gene analysis purely in order to subsequently gain benefits in their insurance contracts in case of a *favorable* test result.[40] As the insured would freely and willingly surrender their genetic information to insurers in such a case, the law provides that even the mere receipt of such information or data about genetic examinations or gene analyses by insurers is prohibited.

Given these reasons for the extension of the scope of protection of § 18 I 1 GenDG to the time after closure of the insurance contract, as well as the prohibition to receive information and data from genetic examinations and analyses already carried out, these provisions are unnecessary for the market of health insurance, although not without a certain – unintended – consequence.

This consequence is the impact of the regulation on the reimbursement possibility of necessary or medically useful genetic examinations and analyses.[41]

In private health insurance, costs which accrue in the course of medical examinations and or treatment of disease are usually paid initially by the insured, who in turn are entitled to recover the expense from their insurer. For this purpose, the medical bill has to contain specific information about the medical examination and treatment undertaken, information which the insurer needs in order to ascertain whether the respective examination or treatment is covered under the individual insurance contract. In case of a hospitalization, this is done through the use of the German Diagnosis Related Groups (G-DRG[42]), which according to §§ 7 I 1 Nr. 1, 9 I 1 Nr. 1 KHEntgG[43] in connection with § 17b KHG[44] have to be used in invoicing. Exactly this information, however, in case of genetic examinations or gene analyses, is now protected by § 18 I 1 GenDG and consequently denied to insurers. Moreover, due to the inclusion of '*data*' about

---

39 § 203 II VVG in connection with §§ 12, 12 a, 12 b I – II a, 12 c German Insurance Supervision Act (Versicherungsaufsichtsgesetz – VAG) and §§ 10, 11, 14 a German Calculation Ordinance (Kalkulationsverordnung – KalV).

40 Bundesregierung, 73.

41 For examples see Ziegler & Ziegler, 42, at footnotes 52 and 53.

42 See for further details: http://www.g-drg.de/cms/G-DRG-System_2012 (last visited July 2012).

43 German Hospital Remuneration Act (Krankenhausentgeltgesetz, BGBl. I S. 1412, 1422, codified as amended in BGBl. I S. 579).

44 German Hospital Financing Act (Krankenhausfinanzierungsgesetz, BGBl. I S. 886).

genetic examinations or gene analyses, insurers are even prohibited from receiving information about the fact (or '*datum*') itself, that a genetic examination or gene analysis has been carried out.[45] In contrast to GINA, which anticipated this problem and consequently stipulates in the 'rules of construction regarding payment' that an insurer is allowed to ask for and use genetic information of an insured for payment purposes[46], the GenDG does not contain such an exemption for payment purposes.

In fact this problem has already penetrated to health service providers who express uncertainty about how invoicing of respective services has to be carried out in the future.[47] Until this problem is solved by the legislator, medical providers bear the risk of uncertainty as to whether a patient himself is able to cover the expenses of such genetic examinations or gene analyses. In this regard, medical providers bear the risk of insolvency of their patients[48], which is a consequence of the GenDG that directly counteracts the core principle of insurance.

d. § 18 I 2 GenDG

Questions also arise in the context of § 18 I 2 GenDG, the obligation of insurance applicants to disclose genetic information to insurers in case of certain insurance types with benefits exceeding the fixed[49] insured sums of EURO 300,000 one-time benefit or EUR 30,000 annual pension.

This provision was enacted as a means to offset adverse selection effects in certain types of insurance where high financial benefits are at stake. The legislator argued that by this provision "the misuse of a knowledge advantage to the detriment of the community of the insured, which is based on the principle of mutual solidarity, shall be prevented".[50] This argumentation by the legislator is astounding,

---

45 Neuhaus, 316; Kröger, 753; Ziegler & Ziegler, 39.

46 See e.g.: Genetic Information Nondiscrimination Act of 2008, Pub. L. No. 110–233, 122 Stat. 881, (2008), § 102 (a)(2), 102 (b)(1)(B).

47 In fact, already the dissemination of genetic information by medical service providers to medical accounting centers is questionable under the GenDG, see e.g.: Gunnar Duttge, et al., Das Gendiagnostikgesetz im Spannungsfeld von Humangenetik und Recht, Univ.-Verl. Göttingen, (2011), 123.

48 Julian Ziegler & Andreas Ziegler, *Zwei-Klassen-Medizin, einmal anders - Konsequenzen des Deutschen Gendiagnostikgesetzes für Privatversicherte*, 17 Forschung & Lehre, (2010), 658–660, 660.

49 Bundesrat, 73; Neuhaus, 316; Kröger, 753; Präve, 860.

50 Bundesrat, 73; Kröger, 753; Wolfram H. Eberbach, *Juristische Aspekte einer individualisierten Medizin*, MedR, (2011), 757–770, 768.

as all types of insurance covered by § 18 I 2 GenDG are not based on the principle of mutual solidarity but, instead, as types of fixed-benefit insurance (or as *Präve* points out: different forms of life insurance[51]), are based on the principle of equivalence of individual risk, individual insurance coverage and individual premium.[52] Obviously and in contrast to the legislator's claim, § 18 I 2 GenDG does not protect the insured within an insurance plan against adverse selection. Rather, the protection is directed at the insurers themselves.[53]

In any case, the insurers, under § 18 I 2 GenDG, are safeguarded against the individual adverse selection potential inherent in an insurance contract, as soon as the limiting insurance sums stated in the law are reached. This means that for any covered type of insurance where the insured amount reaches the fixed stated limits[54] of either EUR 300,000 total benefit or EUR 30,000 annual pension, the protection of § 18 I 1 No. 2 GenDG against the disclosure of genetic health risks does not apply, so that the insurance applicant becomes obliged to disclose all results or data stemming from genetic examinations or gene analyses, if required by the insurer within the underwriting process. This legalizes a claim of insurers which, prior to the enactment of the GenDG was even challenged as being unconstitutional.[55] Moreover, due to the legal determination of fixed insured amounts in § 18 I 2 GenDG, the legislator – incidentally (or accidentally?) – fixed the monetary value of the constitutional right of individual self-determination. The question of whether the protection from genetic discrimination pursuant to § 18 I 1 GenDG applies or not, and therefore the question of whether the right of

---

51 Präve, 860.

52 Bund der Versicherten e.V., Entwurf der Bundesregierung eines Gesetzes über genetische Untersuchungen bei Menschen (Gendiagnostikgesetz – GenDG - BT-Drs. 16/10532) - Stellungnahme, Ausschussdrucksache 16(14)0469(31) Ausschuss für Gesundheit ed., Deutscher Bundestag, (2009), 5; Ziegler & Ziegler, 43–44.

53 This is also conceded by Präve, 860.

54 The questionability of the set limiting insurance sums is discussed by Ziegler & Ziegler, 44, who argue that the basis for the chosen amounts is in itself a questionable voluntary agreement on the use of genetic testing for insurance signed in 2001 by the German Insurance Association (GDV), which violated German anti-trust law.

55 Ausschuss für Forschung Technologie und Technikfolgenabschätzung (20. Ausschuss), Bericht gemäß § 56a der Geschäftsordnung des Deutschen Bundestages zur Technikfolgenabschätzung (TA) BT-Drs. 12/7094 vom 16.03.1994 ed., Deutscher Bundestag, (12. Wahlperiode), BT-Drs. 12/7094 vom 16.03.1994, (1994), 69; Günther Wiese, Genetische Analysen und Rechtsordnung, (1994), 86; Enquete-Kommission „Chancen und Risiken der Gentechnologie", Abschlußbericht BT-Drs. 10/6775 ed., Deutscher Bundestag, (10.Wahlperiode), BT-Drs. 10/6775, (1987), 175; Schöffski, 233.

self-determination supersedes the insurer's interest of protection against adverse selection, might become a question of only one EURO or even only one EURO-cent, depending on the insurance sum agreed at the time of closure of the insurance contract.[56]

Furthermore, the provision of § 18 I 2 GenDG ignores that adverse selection might also occur in insurance types which are not part of the exemption, most likely in health insurance where diseases with high incidence but low mortality can – over the years – also produce costs exceeding the calculated premium coverage.[57] Therefore, the exhaustive enumeration[58] of insurance types in § 18 I 2 GenDG in itself constitutes an unfair discrimination of insured in insurance types not included under this provision if such insurance plans face adverse selection due to the lack of a similar disclosure obligation regarding genetic information.

The exclusion of phenotype analysis

A further problematic issue of the new GenDG arises from the definition of the term "gene analysis" by which phenotype analyses are excluded from the coverage of the law. Although the legislative material stresses that the danger to the individual right of self-determination mainly derives from the novel possibility of analyzing smallest samples in a laboratory[59], this argumentation is not convincing.

Commonly, the analysis of the phenotype of an individual is regarded as the first level of gene analysis.[60] As the phenotype is the result of a combination of genetic and non-genetic factors, genetic traits of an individual may be inferred by a physical examination if the respective characteristics are mainly influenced by the individual's genotype.[61] Apart from already established medical diagnostic methods, such as urine analysis, blood sugar analysis or ultrasonography which all may be highly specific indicators for the presence of a genetic mutation[62], in particular the examination of the individual's medical family history is an important factor of phenotype analysis and allows the calculation of the individual risk to develop a monogenic disorder according to Mendelian rules.[63] Especially in the context

---

56 Ziegler & Ziegler, 44.

57 Id. at 45, with additional references.

58 Kröger, 754.

59 Bundesregierung, 16.

60 Thomas D. Gelehrter & Francis S. Collins, Principles of Medical Genetics, Williams and Wilkins, (1990), 307.

61 Ziegler, 78.

62 Deutsche Forschungsgemeinschaft, 12.

63 Andreas Ziegler, *Basic Mechanism of Monogenic Inheritance*, 40 Epilepsia, (1999), 4–8, 6.

of insurance, phenotype analyses including questionnaires about an insurance applicant's family history are the de facto standard method of risk classification[64] within the underwriting process, as for insurers the source of genetic information is irrelevant insofar as the quality of the information, i.e. the predictive value, permits its use within the risk classification process.[65] In this regard, it also does not seem far-fetched to assume that an insurance applicant would have the same sentiment towards a refusal of insurance coverage due to a genetic risk assessment stemming from an examination of family history as towards a refusal on the basis of a laboratory analysis. As analyses of an individual's family history aim at discovering genetic traits which run in the individual's family and thus in principle may predict the possibility of the individual developing gene-associated diseases in the future, the exclusion of such examinations from the scope of protection of the GenDG means that also under the GenDG, insurers may continue their use of genetic information as a factor in the risk classification process as long as the information is derived from phenotype analysis. This constitutes a severe loophole in the legal protection of insurance applicants against genetic discrimination.

Moreover, the fact that results of family history analyses remain usable in the insurance underwriting process – in connection with § 18 I 1 no. 2 GenDG – leads to worse insurability compared to the legal situation prior to the enactment of the GenDG.[66] This is due to the possibility of insurers basing their decision about concluding an insurance contract with an insurance applicant on genetic information derived from family history analysis while the insurance applicant is, due to § 18 I 1 no. 2 GenDG, banned from – even voluntarily – disclosing results of genetic

---

64 Especially in life insurance questionnaires about the family history are commonly used, see BERBERICH, 281 ; EGON LORENZ, *Zur Berücksichtigung genetischer Tests und ihrer Ergebnisse beim Abschluß von Personenversicherungsverträgen*, VersR, (1999), 1309–1315, 1311; SYBILLE SAHMER, *Private Krankenversicherung und Gentests, in* Genetische Diagnostik und Versicherungsschutz. Die Situation in Deutschland, Felix Thiele (Hrsg.) ed., (2000), 43–49, 47; TAUPITZ, 36.

65 Which is the case if a statistically significant causality between the presence of a disease associated genetic disposition and the development of the respective disease is given, see: ZIEGLER, 78.

66 BRAND, 719; JOHANNES HEYERS, *Prädiktive Gesundheitsinformationen – Persönlichkeitsrechte und Drittinteressen – insbesondere am Beispiel der Gendiagnostik bei Abschluß von Privatversicherungen –*, 27 MedR, (2009), 507–512, 512; ZIEGLER & ZIEGLER, 659; ZIEGLER & ZIEGLER, 38; DIRK LOOSCHELDERS & MÜNSTERISCHER VERSICHERUNGSTAG, Die vorvertragliche Anzeigepflicht des Versicherungsnehmers nach der VVG-Reform: Vorträge, gehalten auf dem 28. Münsterischen Versicherungstag am 20. November 2010, VVW, (2011), 11.

examinations or gene analyses to the insurer.[67] If an insurance applicant had a family history of breast cancer, this information would have to be disclosed to insurers, who would then be allowed to base the risk-calculation on this information. However, if the same person underwent genetic testing of the genes BRCA1 and BRCA2 with a favorable result, i.e. with no more than an average risk of developing breast cancer, insurers would be prohibited from receiving or using that information about the applicant in the risk-calculation, even if the insurance applicant would voluntarily want to disclose the respective result. Consequently the insurance applicant would either face a higher insurance premium or would even be denied insurance based on a misleading result of phenotype analysis despite favorable genotype and the will to disclose this information.[68] This directly contradicts the legislators' intention both to prohibit genetic discrimination in insurance and to safeguard the right of individual self-determination.

5. Comparison of the legal protection: GINA vs. GenDG

As the present comparison of the legal protections offered by GINA and the GenDG shows, both laws try to effectively ban the use of genetic information in the insurance underwriting context but achieve different results.

In contrast to the US American GINA, where the key definition of 'genetic information' covers all four levels of gene analysis and thus provides protection against the discriminatory use of genetic information regardless of its origin, the German GenDG does not contain a definition of 'genetic information'. Instead, the law tries to address the danger of genetic discrimination which the German legislator perceives to be inherent in the respective methods of analysis.

The key definition of 'genetic information' as employed by GINA guarantees the broadest possible basis for the applicability of the law as the term principally covers and thus also protects all potential sources of genetic information. Consequently, GINA offers a comprehensive protection against genetic discrimination by prohibiting the use of any genetic information as long as this information is not about the sex, age or an already manifested disease of the insurance applicant.

In contrast, the German GenDG offers neither comprehensive nor comprehensible protection, the latter because of the multitude of internal contradictions and

67 At the same time this violates the individual's right of self-determination in form of the right to personally decide about disclosure of personal data pursuant to Art. 2 I GG in connection with Art. 1 I 1 GG, see: BVerfGE 65, 1, 43. See also: Ziegler, 131–132 with further references.

68 Ziegler & Ziegler, 38.

exceptions, which become obvious only upon closer scrutiny. The definition framework provided by the law, as well as the lack of a clear definition of the term 'genetic information' create loopholes in the scope of protection from genetic discrimination as well as uncertainty for insured and insurance applicants about their protection from genetic discrimination. Due to the exemption of phenotype analysis, the currently used underwriting process still continues, even if based on questions about genetic information. Thus, genetic discrimination remains possible in this context. In addition, insurance applicants might face unfair insurance conditions despite a favorable genotype or might even be denied insurance coverage on the basis of genetic information derived from 'bad' family history analysis, which is a deterioration of their legal protection compared to the situation prior to the enactment of the GenDG. Last but not least, the disclosure obligation of genetic information in case of high benefits in life insurance and other similar insurance types only benefits insurers and in addition constitutes an unfair discrimination of the insured in insurance types which are not included under this exemption, most notably in health insurance.

6. Conclusion

As the examination of the laws shows, the regulatory approaches towards preventing genetic discrimination differ in the USA and Germany. Due to the much broader scope of applicability of GINA, a much more comprehensive protection against genetic discrimination can be achieved than is the case under the GenDG, which still contains severe loopholes.

References

Andreas Ziegler, Basic Mechanism of Monogenic Inheritance, 40 Epilepsia, (1999), 4–8.

Ausschuß für Forschung Technologie und Technikfolgenabschätzung (20. Ausschuß), Bericht gemäß § 56a der Geschäftsordnung des Deutschen Bundestages zur Technikfolgenabschätzung (TA) BT-Drs. 12/7094 vom 16.03.1994 ed., Deutscher Bundestag, (12. Wahlperiode), BT-Drs. 12/7094 vom 16.03.1994, (1994).

Bund der Versicherten e.V., Entwurf der Bundesregierung eines Gesetzes über genetische Untersuchungen bei Menschen (Gendiagnostikgesetz – GenDG - BT-Drs. 16/10532) - Stellungnahme, Ausschussdrucksache 16(14)0469(31) Ausschuss für Gesundheit ed., Deutscher Bundestag, (2009).

Bundesrat, Gesetzentwurf der Bundesregierung - Entwurf eines Gesetzes über genetische Untersuchungen bei Menschen (Gendiagnostikgesetz - GenDG), Drucksache 633/08 vom 29.08.2008, (2008).

Bundesregierung, Unterrichtung durch die Bundesregierung, Entwurf eines Gesetzes über genetische Untersuchungen bei Menschen (Gendiagnostikgesetz – GenDG) – Drucksache

16/10532 – Gegenäußerung der Bundesregierung zu der Stellungnahme des Bundesrates, Deutscher Bundestag (16. Wahlperiode), BT-Drs. 16/10532 vom 15. 10. 2008, (2008).

Christian Armbrüster, Stellungnahme zum Entwurf eines Gesetzes über genetische Untersuchungen bei Menschen (Gendiagnostikgesetz - GenDG) im Rahmen der Öffentlichen Anhörung des Ausschusses für Gesundheit des Deutschen Bundestages vom 21. Januar 2009, Ausschussdrucksache 16(14)0469(41) Ausschuss für Gesundheit ed., Deutscher Bundestag, (2009).

Deutsche Forschungsgemeinschaft, DFG-Senatskommission - Prädiktive genetische Diagnostik - Wissenschaftliche Grundlagen, praktische Umsetzung und soziale Implementierung (2003), at http://www.dfg.de/aktuelles_presse/reden_stellungnahmen/2003/download/predictive_genetic_diagnosis.pdf,

Dirk Looschelders & Münsterischer Versicherungstag, Die vorvertragliche Anzeigepflicht des Versicherungsnehmers nach der VVG-Reform: Vorträge, gehalten auf dem 28. Münsterischen Versicherungstag am 20. November 2010, VVW, (2011).

Egon Lorenz, Zur Berücksichtigung genetischer Tests und ihrer Ergebnisse beim Abschluß von Personenversicherungsverträgen, VersR, (1999), 1309–1315.

Egon Lorenz, Zur Berücksichtigung genetischer Tests und ihrer Ergebnisse beim Abschluß von Personenversicherungsverträgen - ein Diskussionsbeitrag, in Genetische Diagnostik und Versicherungsschutz. Die Situation in Deutschland, Felix Thiele ed., (2000), 14–35.

Erich R. Prölss & Anton Martin, Versicherungsvertragsgesetz Helmut Kollhosser Ulrich Knappmann, Jürgen Prölss, Wolfgang Voit, Christian Armbrüster ed., C. H. Beck 27 Aufl. ed., (2004).

Erwin Deutsch, Die Genomanalyse: Neue Rechtsprobleme, 19 ZRP, (1986), 1–4.

Erwin Deutsch, Das Persönlichkeitsrecht des Patienten, 192 AcP, (1992), 161–180.

Enquete-Kommission "Chancen und Risiken der Gentechnologie", Abschlußbericht BT-Drs. 10/6775 ed., Deutscher Bundestag, (10.Wahlperiode), BT-Drs. 10/6775, (1987).

George A. Akerlof, The Market for "Lemons": Quality Uncertainty and the Market Mechanism, 84, Quarterly Journal of Economics, (1970), 488–500.

Gunnar Duttge, et al., Das Gendiagnostikgesetz im Spannungsfeld von Humangenetik und Recht, Univ.-Verl. Göttingen, (2011).

Günther Wiese, Genetische Analysen und Rechtsordnung, (1994).

Hanns Bauermeister & Klaus Bohn, Krankenversicherungsmathematik, in Handwörterbuch der Versicherung (HdV), Dieter Farny, et al. eds., (1988), 401–407.

Jessica L. Roberts, Preempting Discrimination: Lessons from the Genetic Information Nondiscrimination Act, 63 Vanderbilt Law Review, (2010), 439–490.

Jessica L. Roberts, The Genetic Information Nondiscrimination Act as an Antidiscrimination Law, 86 Notre Dame Law Review, (2011), 597–648.

Jochen Taupitz, Genetische Diagnostik und Versicherungsrecht, Verlag Versicherungswirtschaft, (2000).

Johannes Heyers, Prädiktive Gesundheitsinformationen – Persönlichkeitsrechte und Drittinteressen – insbesondere am Beispiel der Gendiagnostik bei Abschluß von Privatversicherungen –, 27 MedR, (2009), 507–512.

Julian Ziegler, Genanalysen in der Deutschen Privaten Krankenversicherung, Lang, (2008).

Julian Ziegler, Genetische Diskriminierung in der privaten Krankenversicherung in den USA und Deutschland: Alter Wein in neuen Schläuchen?, 34 DAJV-Newsletter, (2009), 114–122.

Julian Ziegler & Andreas Ziegler, Zwei-Klassen-Medizin, einmal anders - Konsequenzen des Deutschen Gendiagnostikgesetzes für Privatversicherte, 17 Forschung & Lehre, (2010), 658–660.

Julian Ziegler & Andreas Ziegler, Gendiagnostikgesetz und Versicherung: Anspruch und Wirklichkeit, 100 ZVersWiss, (2011), 29–53.

Jürgen Prölss, Der Versicherer als „Treuhänder der Gefahrengemeinschaft" - Zur Wahrnehmung kollektiver Belange der Versicherungsnehmer durch den Privatversicherer, in Festschrift für Karl Larenz zum 80. Geburtstag, (1983), 487–535.

Jürgen Simon, Genomanalyse - Anwendungsmöglichkeiten und rechtlicher Regelungsbedarf, MDR, (1991), 5–14.

Kai-Jochen Neuhaus, Aktuelle Probleme in der Personenversicherung - unter besonderer Berücksichtigung der Berufsunfähigkeitsversicherung -, r + s, (2009), 309–319.

Kerstin Berberich, Zur Zulässigkeit genetischer Tests in der Lebens- und privaten Krankenversicherung, Albrecht, P, Lorenz, E., (1998).

Mark J. Browne, Evidence of adverse selection in the Individual Health Insurance Market, 59 The Journal of Risk and Insurance, (1992), 13–33.

Mark J. Browne & Helene I. Doerpinghaus, Information Asymmetries and Adverse Selection in the Market for Individual Medical Expense Insurance, 60 The Journal of Risk and Insurance, (1993), 300–312.

Michael Spence & Richard Zeckhauser, Insurance, information, and Individual Action, 61 The American Economic Review, (1971), 380–387.

Oliver Brand, Grenzen der vorvertraglichen Anzeigepflichten des Versicherungsnehmers, VersR, (2009), 715–721.

Oliver Schöffski, Genetik und Versicherung aus ökonomischer Sicht, ZVersWiss, (2002), 227–253.

Peter Albrecht, Gesetze der großen Zahlen und Ausgleich im Kollektiv - Bemerkungen zu Grundlagen der Versicherungsproduktion, ZVersWiss, (1982), 501–538.

Peter Präve, Das Recht des Versicherungsnehmers auf gen-informationelle Selbstbestimmung, VersR, (1992), 279–284.

Peter Präve, Das Gendiagnostikgesetz aus versicherungsrechtlicher Sicht, 60 VersR, (2009), 857–912.

R. F. Rich & J. Ziegler, Genetic discrimination in health insurance-comprehensive legal solutions for a (not so) special problem?, 2 Indiana Health Law Rev, (2005), 1–47.

Rainer Paslack & Hilmar Stolte, Gendiagnostik: Stand der Forschung und neue Optionen für Medizin und Wirtschaft, Ministerium für Wirtschaft und Arbeit des Landes Nordrhein-Westfalen, (2003).

Reinhard Damm, Persönlichkeitsschutz und medizintechnische Entwicklung, JZ, (1998), 926-938.

Reinhard Damm, Prädiktive Medizin und Patientenautonomie, MedR, (1999), 437–448.

Sebastian Kröger, Das neue Gendiganostikgesetz und seine Auswirkungen auf den rechtlichen Rahmen beim Abschluss von Versicherungsverträgen, MedR, (2010), 751–756.

Sybille Sahmer, Private Krankenversicherung und Gentests, in Genetische Diagnostik und Versicherungsschutz. Die Situation in Deutschland, Felix Thiele (Hrsg.) ed., (2000), 43–49.

Thomas D. Gelehrter & Francis S. Collins, Principles of Medical Genetics, Williams and Wilkins, (1990).

Wolfram H. Eberbach, Juristische Aspekte einer individualisierten Medizin, MedR, (2011), 757–770.

Enhancement in Medicine and Everyday Life: Ethical Aspects and Requirements for Legal Regulation

Jürgen Simon/Jürgen Robienski

1. Everyday Enhancement as a Social Trend

"Modern society is formatively influenced by increasing pressure and competition. To keep fit in the fight for rare training posts, qualified jobs and promotion chances, to meet the increasing demands of educational institutions and employers, as well as to produce better output than the competition [...] – and nevertheless to save time for the family and hobbies – [...] all these goals need not only permanent fitness and strong self-confidence, but also high resistance and mental flexibility as a psychological basis. An increasing interest will also be aroused in measures and aids which support output from private school lessons, vocational coaching, even wellness methods and psycho-therapeutical aids. Striving for improved efforts is increasingly accompanied by possible ways of raising individually very different levels of efficiency."[1]

In this context, the taking of psycho-active substances has always played an important role: frequent and intensive coffee consumption has advanced to become a common drug with widespread acceptance in the public at large. In the meantime, the arsenal of such available "little helpers" has been enlarged, thanks to immense developments in pharmaceutical research. Even healthy persons take, alongside (hormonal) preparations for the improvement of their physical efficiency und physical wellness, increasingly also special psycho medicines originally developed for therapeutical purposes, using them for the punctual or permanent improvement of their mental (cognitive and emotive) efficiency.[2]

Such a transfer from the medical field to the everyday world of healthy people certainly raises a number of ethical, (socio-) political and, last but not least, legal issues.

---

1 Simon, Jürgen/Robienski, Jürgen/Paslack, Rainer (2008), Enhancement in medicine and every-day-life: A first exploration of ethical implications and legal regulation requirements, Expert opinion for the Office of Technology Assessment by order of the German Parliament, Berlin, p. 3.; also in: Der pharmakologisch verbesserte Mensch. Leistungssteigernde Mittel als gesellschaftliche Herausforderung (ed. Arnold Sauter/Katrin Gerlinger), Studien des Büros für Technikfolgen Abschätzung beim Deutschen Bundestag – 34, Berlin 2012.

2 Simon/Robienski/Paslack (2008), p. 3.

Especially the relatively easy access to "neuro-enhancers" makes everyday enhancement increasingly a topic for medical ethics and technology assessment. The widespread adoption of various anti-depressants (e.g. Prozac) or drugs such as Ritalin, which are also increasingly used by healthy people, can bring about social developments that prepare the basis for a growing readiness to take enhancers. At least the following social trends promote physical and mental enhancement:

- The progressive (self-) medication of increasing parts of society (tablets are more often consumed to "treat" minor feelings of negative health).
- The increasing habituation regarding publicly available enhancement remedies (from energy drinks to cosmetic surgery, but also "classic" coffee, for example), which is accompanied by a corresponding trivialization of their consumption.
- A growing interest in the wealth of health and fitness products (including "functional foods") and the relatively high prevalence of enhancers in popular sport (especially in the bodybuilding community).
- A gradually changing picture in the public eye of what has to be treated as an illness, linked with an expansion of the term illness itself – often not covered by medical terminology – to include all sorts of disorders and behavior perceived as "not normal".
- The increasing pressure to perform and compete in terms of professional career opportunities and job security (already noticeable in parents' homes).
- The broad application of everyday enhancement remedies and promotional propagation of new "diseases" requiring treatment, as well as supposedly riskier standard deviations by the "enhancement industry".
- The ever more diverse and intense R&D in areas such as brain research (e.g. neuronal receptor research), the medical treatment of mental illnesses and disorders, and future "genetic enhancement" (in the wake of the development of gene therapy).[3]

In summary: the common striving to dominate our body by manufacturing is a general characteristic of what we would describe as "Promethean modernity" or a sign of the "Promethean human being". For the Promethean man, the possible is considered as the true reality.[4] It aims at the realization of technological projects that do not stop at one's own constitution – the "survival of the fittest" in this case

3 Simon/Robienski/Paslack (2008), p. 4.
4 Simon/Robienski/Paslack (2008), p. 8.

includes the "survival of the prettiest", or as Caysa says: "The beautiful body becomes the object of the final possible realization of their dream, the desired 'real' self."[5] Here, the principally unscrupulous move to handle body enhancement – or doping – is close because the most modern forms of doping are nothing other than the leading products of the body industry, which are now becoming democratized. The doped athlete appears in this context only as a harbinger of a technology-obsessed body avant-garde who has finally overcome the shortcomings of the natural body (with the old authentic body) by retrofitting and improving it with high technology.[6] The idea of the Promethean man is always based on the adoption of a naturally "deficient man".

Enhancement is therefore only a tried and tested way to remove its natural deficiencies (the "mistake of nature"). And, last but not least, democratized enhancement is able to contribute to the "equalization of happiness" (Caysa). A utilitarian idea that neurologist Linke already had in mind when he wrote in 1996: "And why would you stop at the brain bringing happiness? What are the limits of intervention in nature? Should we give up umbrellas and dikes because they oppose the natural flow of water? Should we refrain from brain surgery, because it alters natural personalities? Why should we not model out individualities from a given brain so that it is not only corresponding to our understanding of health, but perhaps also to the concept of happiness?"[7]

Anyway, it seems to be the case that enhancement (or doping) "anticipates the new body normality of the biotechnological age" (Caysa). Industrialization of the body is apparently in line with a development which has always been active in the West and has forced domestication and rationalization of the human body in modern times: submission to the means-to-an-end or even cost-benefit calculus. But this capitalization of the body of individuals in the form of freedom of body use, free control over one's own body and the individual utility value of the body is perceived generally in a positive light.[8] Associated with the current body commodification is a change in direction of a loss of individual body diversity as a result of serialization and standardization through the arising of "body modes" (especially based on ideal body cosmetic surgery, as well as on the smoothing of

---

5 Caysa, Volker (2008), Doping und das Problem des Maßes einer „natürlichen" Eigenleistung des Menschen, in: Latzel, Klaus/Niedhammer, Lutz (ed.), Hormone und Hochleistung – Doping in Ost und West, Köln/Weimar/Wien, pp. 245 ff. (248).

6 Caysa (2008), p. 249.

7 Linke, Detlef (1996), Hirnverpflanzung. Die erste Unsterblichkeit auf Erden, Reinbek bei Hamburg, p. 149 f.

8 Caysa 2008, p. 251.

moods and the normalizing "balancing" of alleged cognitive deficits. This leads ultimately to a "McDonaldization" of the body world - and not only that).[9]

2. Legal and Regulatory Aspects of Human Enhancement

Given biomedical, pharmacological and neurological progress, there are new enhancement options that go far beyond mere cosmetic surgery and the like. This enhancement affects the whole of society and cultural developments which also require an appraisal and response of the legislature as a regulatory power. To this extent Kirchhof states correctly: "Law must take care of medical progress, promote medical intellectual curiosity, but also moderate in order to protect the people."[10]

Several of these measures, such as cosmetic surgery, have already been the subject of higher court jurisdiction and jurisprudential debate in Germany. These measures have been discussed, however, only in isolation. A jurisprudential discussion of enhancement as a whole has not yet taken place.[11]

Jurisprudence is apparently lagging behind the social development of enhancement. Not all enhancement-related issues have been addressed by legislation to date. On the contrary, the concept of disease, for example, is still defined legally in a way which does not sufficiently take into consideration the new possibilities and consequences of enhancement. Also, in jurisprudence experts can look for a comprehensive and differentiated discourse in vain. Although the discussion has been opened, it is still in its early stages and seeking to identify the related problem areas. But in the last decade there have been some legal investigations. Very interesting is a study by the Science and Technology Options Assessment (STOA) of the European Parliament, the Human Enhancement Study of 2009.[12] With it we can summarize for the German, as well as the European, level that many enhancement technologies have been developed in a medical context. They can also be used beyond medical practice. Especially on the international level, the notion of applications in the sense of enhancement has been discussed in a broader way. Thus, in spite of being quite a recent phenomenon in the intensity of its scientific

9 Simon/Robienski/Paslack (2008), p. 9.

10 Thorbrietz, Petra (2008), Süddeutsche Zeitung Wissen. Dezember 2008, p. 21 f.

11 Eberbach, Wolfram (2009), Die Verbesserung des Menschen. Tatsächliche und rechtliche Aspekte der wunscherfüllenden Medizin, in: Der neue Mensch? Enhancement und Genetik (Nikolaus Knoepffler/Julian Savalescu ed.), Freiburg/München, pp. 213 ff.

12 Coenen, Christopher et al. (2009), Human Enhancement Study, European Parliament, Science and Technology Options Assessment (STOA), pp. 134 ff.

research and it not being put into practice very often yet, a discussion of the legal questions raised by enhancement technologies has at least begun.

Furthermore, the needs or options for juridical regulations have been discussed in some studies for fields of technology or application that could bear a potential for enhancing or altering human functions.[13] In special areas such as human genetics, or pharmaceuticals and medicine in general, governments have put legal regulations into force. Thus, good (medical) practice and the rights of doctors and patients are not only intensively discussed, but regulated. Ethical principles are used to prevent the misuse of biomedical applications. Current regulations, however, do not cover or deal with a central problem connected with enhancement technologies. This is the distinction between the medical use of devices, technologies and procedures, on the one hand, and, on the other hand, the "non-medical" use of the same technologies for enhancing bodily functions, the capacities of the human mind, or manipulating mental states, be it for lifestyle, daily life or professional purposes. As an expert discussion of this has not even begun yet, the abovementioned Human Enhancement Study of 2009 reports that it can only give some indications and ideas of the juridical problems that might arise from enhancement technologies.[14]

The basic and most obvious practical juridical problem caused by enhancement technologies is associated with the problem of separating a medical application of a given technology from a non-therapeutic enhancement application and developing a clear distinction between legally allowed (or even obligatory) use and ethically doubtful practice that consequently would be liable to legal regulation or prosecution.[15] As has been argued, it is difficult to separate an illegitimate from a legitimate use of enhancement technologies. Neither of the distinctions under discussion – natural vs. unnatural or therapeutic use vs. non-therapeutic – is sufficiently selective.

An illustration of the problems involved in clearly distinguishing harmful or ethically questionable "illegal" uses from beneficial and ethically acceptable "legal" uses of enhancement technologies is the current practice of prosecution for doping in sports. The problem is not only how to keep pace with the development of ever-new drugs or other methods for enhancing bodily performance, and how to

---

13 Eberbach (2009), pp. 213 ff.; Monbaron, Samuel/ Sprumont, Dominique (2010), Le developpement humain artificiel communement appele "human enhancement". Legislation actuelle et besoins de reglementation, Neuchatel; Eckardt, Anne et al. (2011), Human Enhancement, TA-Swiss 56/2011, Zürich, pp. 192 ff.

14 Simon/Robienski/Paslack (2008), p. 13. 13 Greely, Hank (2004), Disabilities, Enhancement and the Meanings of Sport, in: Stanford Public Law Working Paper, No. 82.

15 Coenen et al. (2009), pp. 134 ff.

prove in each individual case that misuse is present; the problem is also to justify a ban on certain substances and practices while allowing others. As Greely has shown, it can easily be argued that the current practice of prosecuting for doping lacks logical consistency and juridical justification.[16] In sports, all kinds of practices to increase physical performance (fortitude, endurance, readiness and speed) are allowed, despite the fact that many of them involve health risks, e.g. physical conditioning, special nutrition and nutritional supplements, or surgery to avoid injuries.

Doping is only one decisive issue, and it is not sufficient for a clear definition of permitted enhancements in other areas. Therefore, it is very helpful to try to find a definition of enhancement. Basically, medical treatment for the prevention or treatment of an objectively present disease or disability is necessary in the legal sense. Measures of enhancement, however, are performed solely due to the desire of an individual and his subjective feeling. Whether this legal distinction in this rather unique form can be maintained in the future is, however, questionable in view of the progressive development of society.

2.1 Constitutional Issues in Germany

Further development of enhancement is likely to bring great alterations in the appearance of human beings.[17] Biomedical progress promises a variety of ways to design one's own external and internal appearance. In particular, genetic enhancement is expected to project a bio-technological, mechanistic view of the human body. The author Charlotte Kerner writes: "The belief in the body as a natural product is shaken".[18]

The human being can be a comprehensive designer of himself and his descendants. We may to a certain extent already be the designers of ourselves by using the methods of plastic surgery. Also, a selection of our own offspring by means of certain already created genetic traits is possible. The active shaping of our own offspring by "surgical" interventions in the genes, however, is still a vision.[19] In

---

16 Simon/Robienski/Paslack (2008), pp. 12 f.

17 Greely, Hank (2004), op. cit.

18 As Savulescu, Julian (2006) writes in his intensively noticed article, 'Genetic Interventions and the Ethics of Enhancement of Human Beings', in: B. Steinbock (ed.) The Oxford Handbook of Bioethics, Oxford, pp. 516 ff., enhancement will be necessary to overcome the challenges of this century. See also Ranisch, Robert/Savalescu, Julian (2007), Ethik und Enhancement, in: Der neue Mensch? Enhancement und Genetik (ed. Nikolaus Knoepffler/ Julian Savalescu), Freiburg/München, pp. 21 ff.

19 Kerner, Charlotte (2008), Wann wird der erste Kopf verpflanzt? in: Bild der Wissenschaft 12/2008, p. 29 ff. (30).

this regard, the question arises of whether the implementation of such opportunities is compatible even with the constitutional order, the principle of human dignity under Article 1 GG and the "rights of others". Are we allowed to be a constructive and selective creator and designer of our own selves, or even of our descendants, or are we setting ourselves up to be the mere objects of a biomedical-industrial complex, and if so, is the protection of the state needed?

Basically, the measures to improve one's body through cosmetic surgery or the like, which are carried out without a need for treatment, given to a person at his request, are an expression of his constitutionally guaranteed individual right to self-determination under Article 2, para 1 German Constitution (GC).[20]

The free development of human personality and dignity are the supreme value of the constitutional law. With the right of free development of the personality, the core area of private life is protected. This includes especially the design and presentation of one's own physical appearance. This right is limited only by the constitutional order, the rights of others and the moral laws. A breach of these barriers is not seen in the body improvement options given so far.

But this raises the question of whether the extent of the self-determination right also includes a right of freedom and design of genetic disposition if the genetic predisposition of the human body due to biomedical possibilities is no longer axiomatic and immutable.[21] Also, the genetic disposition is presented as the core area of privacy, as a last inviolable sphere of human freedom, in which interventions are inadmissible.[22]

On the other hand, there follows from Article 2 para 1 and Article 2 para 2 1 GC an objective legal obligation of the state to protect the right of life and physical integrity.[23] As enhancement, especially genetic enhancement, can result in significant, unanticipated adverse consequences for health (this is already shown by the number of unsuccessful cosmetic surgeries), the question arises as to whether the state is obliged to protect people against these measures. This is especially true with regard to minors. It has been noted that increasingly minors, at the request of their parents or of their own volition but with the consent of their parents, are having cosmetic surgery performed.[24]

---

20 Diedrich, H./Schön, F. (2008), Wer sind wir in 20 Jahren? In: Welt der Wunder 7/08, p. 46.

21 Simon/Robienski/Paslack (2008), p. 16.

22 Beck, Susanne (2006), Enhancement – die fehlende rechtliche Debatte einer gesellschaftlichen Entwicklung, in: MedR, p. 98.

23 Entscheidungen des deutschen Bundesverfassungsgerichts (BVerfGE) 6, 32, 41.

24 BVerfGE 85, 191, 212; 88, 203, 251; 90, 145, 195

With appropriate measures, such as the prohibition of certain methods of enhancement, the state would potentially interfere in the fundamental right to freedom of occupation of providers of such measures, so that a violation of Article 12 GC seems possible.

If genetic enhancement in living humans is possible and allowed, the question arises in turn of whether a total prohibition of pre-implantation (PGD) or a restricted one, as in Germany, can now be maintained. This applies even more since its prohibition is not regular in other countries of the European Community, and given the dynamics of scientific progress it is only a matter of time until legal obstacles fall.[25]

Relating to pre-implantation and prenatal diagnosis, the question of the human image arises to a certain degree. When will a life be seen as unworthy due to its genetic disposition, so that it may be "discarded" or "modified"? When does an embryo begin to be a human being, a person, and fall within the protection of Articles 1, 2 GC?

If the option exists of designing one's own self or offspring, the question also arises of when "an illegal body or mental state deviating from the role model of a healthy person" is given within the meaning of the disease concept, which "requires medical treatment". Is the developing life which has, for example, a genetic predisposition to a little below average body size, but not pathologically, entitled to "treatment" if this were determined in the context of prenatal diagnosis and could still be treated in the womb?[26]

The fact that these issues are very realistic is shown by a recent example from England. There a deaf couple who already had a deaf child wanted to conceive a second child using IVF, and for this purpose an embryo should be selected by IVF which definitely also had genetic deafness. The couple did not see themselves as having a disability, but rather as having their "own way" with their own language. This is a view which is not rarely encountered in deaf people.

Moreover, it is feared that the further unbridled spread of enhancement may lead to enhancement measures that may be undertaken to escape an otherwise to be feared stigmatization and discrimination, such as because of skin color or other racial characteristics. The pop star, Michael Jackson, represents an extreme example of such behavior. Enhancement can therefore contribute to a worsening of the problem of discrimination in society. This raises the question of whether the state is, therefore, forced to act to protect the individual?

---

25 Thorbrietz (2008), p. 22; Simon/Robienski/Paslack (2008), p. 17.

26 Eberbach 2008, p. 331.

2.2 General Problems

In this context, there are also more biomedical procedures to investigate, such as xenotransplantation, therapeutic cloning and tissue engineering, but also other new technological processes, such as the "green gene technology" (and the associated genetically modified foods) or nanotechnology, regarding their relevance for enhancement and legal issues occurring in this respect.

Enhancement is a classic example of specific challenges that the (bio) technological progress creates for legislature. Biotechnology, or more specifically biomedicine is highly innovative and has emerged with a panache that is unparalleled. In order not to hinder the research and desired positive progress, the legislature is forced to provide "innovative law", that means to create not only a legal framework to regulate future developments, but also to provide justice with it.[27]

At the same time, it is imperative that arrangements will be made that include a high level of sustainability regarding the undeniable risks in order to protect future generations. Particularly genetic enhancement harbors the danger that use of these technologies and their effects are irreversible. It should be noted that basic biotechnological research is still far from answering all questions. For a long time it will not be fully understood how genetic processes work, which impact genetic manipulations have or, in particular, which risks they pose. This applies to both the biological and the sociological consequences.[28]

3. Summary

As we can see, the direct application of German constitutional law throws up more questions than answers, and this will be the same for other legal issues in Germany and other fields like criminal law, medical law or insurance law. "The fundamental legal principle of liberal societies, the right of self determination and autonomy of the individual, does not give a clear guideline about restricting human enhancement as long as an individual decision for making use of enhancement technologies does not interfere with the fundamental rights of others."[29]

The question is how the legislature should differentiate between ethically acceptable and unacceptable forms of enhancement. A first step could be the concept of the "person" to inform the regulatory discourse. "In this respect it is

27 Simon/Robienski/Paslack (2008), p. 18.
28 Simon/Robienski/Paslack (2008), p. 31.
29 Simon/Robienski/Paslack (2008), p. 31.

not only the clear impairment of a person's autonomy that may be rejected, as in the case of drugs that directly affect a person's capacity for self-determination in the sense of taking accountable decisions. It might also be asked to what extent the diffusion and application of enhancement technologies might change the social context and culture in a way that the extent to which an autonomous decision for or against making use of enhancement technologies is possible at all might be doubted. As soon as enhancement technologies have become a 'normal' aspect of competition in everyday professional life, it appears to be problematic to support a free and sovereign decision of an individual to apply enhancement."[30]

Very important, following the approach of Savalescu, is also the question of the extent to which enhancement technologies may impair the personal structure of an individual in a new, perhaps ethically questionable but necessary way to solve the problems of this century.[31] It is right that personal autonomy implies not only personal sovereignty (self-government) but also "authenticity" in the sense of regarding one's own action and behavior as belonging and being in line with one's own "identity".[32] But it has urgently to be discussed whether enhancement technologies that intervene in a person's emotional state by manipulating his or her ability to suppress certain emotions and anxieties (as used, e.g. by military pilots), as well as enhancement technologies to "steer" or "filter" a person's ability to remember parts of his personal history, could – when exceeding beyond "healthy" barriers – be regarded as impairing or even destroying (at least temporarily) the identity of a person, and thus affecting his/her autonomy by inducing action and behavior that can be regarded or experienced as being not an "integrated" part of a person's identity.[33] If we follow Ranisch and Savalescu we can not only state that enhancement may improve chances for a better life, but also that human enhancement could be one necessary means for facing the challenges of the 21st century. The provocative thesis is that we should use all possible means to develop our person and personality as far as possible – including with enhancement.

30 Coenen, Christopher et al. (2009), p. 138.
31 Coenen, Christopher et al. (2009), p. 139.
32 Savulescu, Julian (2006), p. 516 ff.
33 Coenen, Christopher et al. (2009), p. 139.

References

Beck, Susanne (2006): Enhancement – Die fehlende rechtliche Debatte einer gesellschaftlichen Entwicklung. In: *MedR*, pp. 95–102.

Caysa, Volker (2008): Doping und das Problem des Maßes einer „natürlichen" Eigenleistung des Menschen, in: Latzel, Klaus/Niedhammer, Lutz (ed.), Hormone und Hochleistung – Doping in Ost und West, Köln/Weimar/Wien, pp. 245–261.

Coenen, Christopher et al. (2009): *Human Enhancement Study, European Parliament, Science and Technology Options Assessment (STOA).*

Diedrich, H./Schön, F. (2008): Wer sind wir in 20 Jahren? In: *Welt der Wunder* 7/08, p. 46.

Eberbach, Wolfram (2009): Die Verbesserung des Menschen. Tatsächliche und rechtliche Aspekte der wunscherfüllenden Medizin. In: Nikolaus Knoepffler/Julian Savalescu (eds.): *Der neue Mensch? Enhancement und Genetik*, Freiburg/München, pp. 213–250.

Eckardt, Anne et al. (2011) : *Human Enhancement*. TA-Swiss 56/2011, Zürich.

Greely, Hank (2004): Disabilities, Enhancement and the Meanings of Sport. In: *Stanford Public Law Working Paper*, No. 82.

Kerner, Charlotte (2008): Wann wird der erste Kopf verpflanzt? In: *Bild der Wissenschaft*, 12/2008, p. 28–31.

Linke, Detlef (1996): *Hirnverpflanzung. Die erste Unsterblichkeit auf Erden*. Reinbek bei Hamburg.

Monbaron, Samuel/Sprumont, Dominique (2010) : *Le developpement humain artificiel communement appele „human enhancement". Legislation actuelle et besoins de reglementation.* Neuchatel.

Ranisch, Robert/Savalescu, Julian (2007): Ethik und Enhancement. In: Nikolaus Knoepffler/ Julian Savalescu (eds.): *Der neue Mensch? Enhancement und Genetik*, Freiburg/München, Freiburg/München, pp. 21–53.

Savulescu, Julian (2006): Genetic Interventions and the Ethics of Enhancement of Human Beings. In: B. Steinbock (ed.) *The Oxford Handbook of Bioethics*, Oxford, pp. 516–535.

Simon, Jürgen/Robienski, Jürgen/Paslack, Rainer (2008): *Enhancement in medicine and everyday-life: A first exploration of ethical implications and legal regulation requirements, Expert opinion for the Office of Technology Assessment by order of the German Parliament*, Berlin; also in: Arnold Sauter /Katrin Gerlinger (ed.): *Der pharmakologisch verbesserte Mensch. Leistungssteigernde Mittel als gesellschaftliche Herausforderung*. Studien des Büros für Technikfolgen Abschätzung beim Deutschen Bundestag – 34, Berlin 2012.

Thorbrietz, Petra (2008), *Süddeutsche Zeitung Wissen*. Dezember 2008, p. 21.

Data Mining of Medical Data: Opportunities and Challenges in Mining Association Rules

Dan A. Simovici

> *The beginning clinical clerk, the house officer and the practicing physician are all confronted with conditions that are frustrating in every phase of medical action.* [...] *To deal effectively with these frustrations it will be necessary to develop a more organized approach to the medical record, a more rational acceptance and use of the paramedical personnel and a more positive attitude about the computer in medicine.*
>
> (L. L. Weed: Medical records that guide and teach, New England Journal of Medicine, 1968)

1. Introduction

Data Mining (DM) is the process that discovers new patterns embedded in large data sets. DM makes use of this information to build predictive models. DM is grounded in artificial intelligence, databases, and statistics.

The health care industry requires the use of DM because of it generates huge and complex volumes of data. Thus, un-automated analysis has become both expensive and impractical. The existence of insurance fraud and abuse impels insurers to use DM. DM can generate information that can be useful to all stakeholders in health care, including patients by identifying effective treatments and best practices.

DM came into prominence in mid 90s because computers made possible the fast construction of huge data warehouses, containing potentially large amounts of information. The modern day statistical techniques and the advances in probability theory offered the necessary analytical tools.

The history of data and its contents is much older. Huge collections of data were built over hundreds and thousands of years by various forms of government and scientists. A famous case is the vast collection of very accurate planetary observations of the Danish astronomer Tycho Brahe (Dec. 14, 1546, Knutstorp Castle – Oct. 24, 1601, Prague). The knowledge embedded in this data – the laws of the movements of the planets were discovered by his successor Johannes Kepler (Dec. 27, 1571, Weil der Stadt – Nov. 15, 1630, Regensburg) and were confirmed by the work of Newton.

The main DM activities consist of description and visualization, seeking associations between data elements, grouping data into sets of similar records (a process known as clustering), data classification, prediction based on trends that can be extracted from data, etc.

DM applications in health care are numerous and already well established: evaluating treatment effectiveness, health care management, the analysis of relationships between patients and providers of care, pharmacovigilance, fraud and abuse detection. Despite the obvious benefits, there exist many limitations and difficulties in adapting DM analysis techniques.

DM can be limited by the accessibility to data that often is distributed in different settings (clinical, administrative, insurers, labs, etc.). Data may be incomplete, corrupted, noisy, or inconsistent. There exist ethical, legal and social issues (data ownership, privacy concerns).

Many patterns find in DM may be the result of random fluctuations, so many such patterns may be useless, which requires a serious statistical analysis. DM of medical data requires specific medical knowledge as well as knowledge of DM technology and, last but not least, DM requires institutional commitment and funding.

In this survey we will focus on exploring data in the pursuit of association rules, a concept that formalizes probabilistic dependencies between parts of data. We begin by introducing the notion of *table*, the main data structure involved in this process. Then, we introduce formally association rules and their main parameters, support and confidence.

We discuss the use of AR in the study of nosocomial infections, adverse drug reactions and issues related to the lack of transitivity of association rules which are relevant for medical applications. To avoid interruptions in the flow of ideas of the paper we relegated the technicalities in an appendix.

2. Tables and Relational Databases

A *table* is an aggregate that consists of

- a *table name*;
- a *heading* that consists of a set of symbols called attributes;
- a *content* that is a multiset of rows: we could have multiple copies of the same row.

Each attribute A has attached to it a *domain dom(A)*, a set that contains at least two elements.

A *row* is a sequence of values $(a_1, a_2, \ldots, a_n)$ such that a_i is a member of dom (A_i) for $1 \leq i \leq n$.

Graphically, a table looks exactly as our intuitive concept would suggest (as we show in the next example.

Example 1:

OBJECTS

	shape	Length	Width	height	color
1	cube	5	5	5	red
2	sphere	3	3	3	blue
3	pyramid	5	6	4	blue
4	cube	2	2	2	Red
5	sphere	3	3	3	Blue

The heading of the table OBJECTS consists of 5 attributes. The domains of attributes *shape* and *color* are

dom(shape) = {cube, sphere, pyramid},
dom(color) = {red, blue}

have no natural ordering. Such domains are said to be *categorial.*

The domains of the remaining attributes, length, width, height are numerical. – A special role is played by *binary tables* due to their capabilities for representing collections of sets. In such tables the domain of every attribute is the set {0,1} and every tuple is a sequence of 0s and 1s.

Let $S = \{s_1, \ldots, s_n\}$ be a set and let T be a subset of S. This subset can be represented by a sequence $(t_1, \ldots, t_n)$, where $t_i = 1$ if t_i is a member of T and $t_i = 0$, otherwise. Binary tables were used in analyzing purchase patterns of supermarket customers, documents containing words, etc. In the initial literature dealing with frequent item sets and association rules (Agrawal, T. Imielinski & Swami, 1993; Agrawal & Schaffer, Parallel mining of association rules, 1996) the goal of this analysis was to determine what items people buy together, regardless of who they are.

Example 2:
A fictional convenience store sells milk, bread, butter, beer, and diapers. The purchase records of seven customers are listed below.

Content of basket

1 {milk, bread, butter, diapers}
2 {bread, beer, diapers}

3 {milk, bread, butter, beer}
4 {bread, butter, diapers}
5 {milk, butter, beer, diapers}
6 {milk, butter}
7 {butter, beer}

and represented in the following binary table:

	milk	Bread	butter	beer	diapers
t_1	1	1	1	0	1
t_2	0	1	0	1	1
t_3	1	1	1	1	0
t_4	0	1	1	0	1
t_5	1	0	1	1	1
t_6	1	0	1	0	0
t_7	0	0	1	1	0

This table indicates, for example, that the fourth customer bought milk, bread, beer, and diapers.

The tabular representation of collections of sets facilitates the introduction of the notion of support of an item set. If X is an item set, the *support* of X is the number of tuples that have 1s in all positions that correspond to the attributes of X. Equivalently, this is the number of baskets that contain X.

Example 3:
For the transaction set defined in Example 2 we have

supp(milk) = 4	supp(bread) = 4
supp(milk bread) = 2	supp(milk bread butter) = 2

Note that the larger the attribute set, the smaller the support: if X is a subset of Y then $supp(Y) \leq supp(X)$. The number

$$\frac{supp(X)}{N}$$

estimates the probability that a randomly chosen transaction contains all elements of X, where N is the total number of transactions. The support of an item set is expressed fractionally (as above) or in percentages. For example, the relative value of the support of the set *milk bread* is 2/7 or 28.57%.

3. Association Rules as Knowledge Embedded in Data

An *association rule* (AR) is a pair (X, Y) of sets of attributes, denoted by $X \rightarrow Y$. X is the antecedent and Y is the consequent of the rule $X \rightarrow Y$.

The simplest parameters associated to an association rule are its *support* and *confidence.*

The *support of a rule* $X \rightarrow Y$ is the number of records that contain all items of, X

$$supp(X \rightarrow Y) = supp(X).$$

The *confidence* of $X \rightarrow Y$ is the number

$$conf(X \rightarrow Y) = \frac{sipp(XY)}{supp(X)}.$$

Clearly, the confidence of $X \rightarrow Y$ is an estimation of the probability that a record that contains the items of X, chosen at random, will contain the items of Y.

Example 4:
For the table

A B C D R
t_1 1 1 0 0 0
t_2 1 0 1 0 0
t_3 0 0 0 1 0
t_4 1 1 1 0 0
t_5 1 0 0 0 1
t_6 1 0 0 0 1
t_7 1 1 0 0 0

and the association rule $AB \rightarrow C$, support equals 3 and confidence is

$$conf(AB \rightarrow C) = \frac{supp(ABC)}{supp(AB)} = \frac{1}{3} = 0.33$$

An AR $X \rightarrow Y$ *holds with support* μ *and* c *confidence* if and $supp(XY) \geq \mu$ *and* $conf(X \rightarrow Y) \geq c$.

Association rules of the form $X \rightarrow Y$, where Y is included in X are called *vacuous* because $conf(X \rightarrow Y) = \dfrac{supp(XY)}{supp(Y)} = 1$ regardless of the actual data set; such rules are not informative and they are also referred to as trivial rules.

Despite its simple formulation, finding association rules with a prescribed support and confidence can offer formidable computational challenges.

For data sets having n attributes there are $3^n - 2^n$$ possible association rules (see, for example (Simovici & Djeraba, 2008)). Even for modest values of n the number of possible nontrivial association rules is very large and the number of collection of possible rules is immense. For example, $n = 20$ for there exist more that one billion non-trivial AR and more than $10^{300000000}$ sets of AR (for comparison, there are 10^{80} atoms in the known universe!). Thus, a considerable effort in DM has been invested in designing efficient algorithms for computing association rules embedded in data sets.

To find an AR $X \to Y$ with support μ and confidence n we need to:

- find an item set U that is at least μ-frequent, that is, $supp(U) \geq \mu$;
- find a subset V of U such that $supp(V) \leq \frac{supp(U)}{c}$

The item sets U and V define the AR $X \to Y_c$ where $X = U - V$ and $Y = V$, having support at least equal to μ and confidence at least equal to c. Thus, computing association rule amounts to computing frequent item sets.

The most common algorithm for computing frequent item sets is the *Apriori algorithm* obtained by Agarwal, Imielinski and Swami (Agarwal, Imielinski & Swami, 1993), which consists of the following main steps:

- detect all items that have a support at least equal to μ;
- for successive numbers $i \geq 2$ join item sets that contain i items with individual item sets (candidate generation phase);
- evaluate the resulting item sets and retain only those who have sufficient support (evaluation phase).

Without entering details, we mention that the algorithm raises non-trivial issues of memory management because often large data sets cannot be accommodated entirely in the main memory of computers. There are numerous references on Apriori implementations that examine these problems (Steinbach, Karypis & Kumar, 2000; Adamo, 2001; Zaki & Hsiao, 2005; Han, Pei, & Yin, 2000; Agarwal, Agarwal & Prasad, 2001).

4. Association Rules and Nosocomial Infections

The study of development of drug resistance of bacteria involved in intra-hospital infections has been pursued in (Brosetts & Hymel, 2008; Brosetts, Sprague, Hardin, Waites, Jones & Moser, 1998) and many other reports.

Among the Gram-negative bacteria which are notorious for their drug resistance, Pseudomonas aeruginosa is a common cause of infections in humans and its transmission is caused by medical equipment, including catheters.

The data collection includes records that describe single *Pseudomonas aeruginosa* isolates. The attributes of the records are:

1. date reported;
2. source of isolate (sputum, blood);
3. location of patient in the hospital;
4. patient's home zip code;
5. resistant (R), intermediate resistance (I), susceptible (S) for piperacillin, ticarcillin/clavulanate, ceftazidime, imipenem, amikacin, gentamicine, tobramycine, ciprofloxacin.

Records passed through a pre-processing phase, when duplicate records were removed, so each patient had one isolate per month. The system was designed to detect patterns of increasing resistance to antimicrobials; therefore, items of the form S-antimicrobial were removed.

Data was partitioned horizontally in time-slices; in each slice identification of association rules with high support is performed and the confidence of these rules was computed. Then, the variation of the confidence of a rule $X \rightarrow Y$ between the current time-slice and the confidence of the same rule in previous time slices is calculated. If a substantial increase in the confidence occurs (as verified using an appropriate statistical test) relative to the previous partition(s), this finding constitutes an *event*.

More specifically, data was partitioned horizontally in

a. 12 one-month fragments: 2,000 ARs;
b. 4 three-month fragments: 12,000 ARs;
c. 2 six-month fragments: 20,000 ARs.

Minimum support for an item was 2 and minimum support for an AR was 10.

The investigators sought short-lived interesting patterns in slices of type A, and long-lived interesting patterns in slices of type C.

Note that for AR of the form $\emptyset \rightarrow Y$ we have $supp(\emptyset \rightarrow Y) = supp(\emptyset) = n$ and $conf(\ \rightarrow Y)\dfrac{supp(Y)}{n}$, which shows that only confidence is significant and equals the probability of Y.

Thus, $conf(R - antimicrobial)$ gives the probability that *Pseudomonas aeruginosa* develops resistance to the antimicrobial; variation in the level of confidence are evaluated on a monthly, quarterly, and semestrial basis.

The selection of association rules was based on the variance in their confidence as follows:

1. For each AR $X \rightarrow Y$ the confidence in P_c, $conf(X \rightarrow Y, P_c)$ was compared with $conf(X \rightarrow Y, P_d)$, the confidence of $X \rightarrow Y$ in the last data set P_d in which $X \rightarrow Y$ was found which precedes P_c.
2. The comparison of confidences is done using a χ^2-comparison of two proportions, or when the number of expected value is small, by the Fisher exact test.
3. If $conf(X \rightarrow Y, P_c) \geq conf(X \rightarrow Y, P_d)$ and the probability that the difference between the proportions occurred by chance is less than 5%, then this finding was presented to the user.

Among the AR found in are the following:

Ø → R – ticarcillin/clavulanate R – ceftazidime R – piperacillin

[A jump from 4% (Oct) to 8% (Nov) to 11% (Dec) suggests that the isolate is resistant to ticarcillin/clavulanate, ceftazidime and piperacillin]

R – ceftazidime R – piperacillin → sputum R – ticarcillin/clavulanate

[A jump from 8% (Feb) to 32% (Aug); it is likely that the isolate is from sputum and is ticarcillin resistent *given that* is resistant to ceftazidime and piperacillin]

R – piperacillin → sputum R – ticarcillin/clavulanate R – ceftazidime

[An increase from 6% (Q3) to 26% (Q4) in the probability that the isolate is from sputum, is ticarcillin/clavulanate and ceftazidime resistant *given that* is piperacillin resistant]

R – ticarcillin/clavulanate → sputum R – ceftazidime R – piperacillin

[An increase from 7% (Q3) to 24% (Q4) in the probability that isolate is from sputum, is ceftazidime and piperacillin resistant *given that* is ticarcilline/clavulanate resistant]

R – ticarcillin/clavulanate R – ceftazidime R – piperacillin\lw sputum

[An increase from 12% (Q3) to 42% (Q4) in the probability that the isolate is from sputum given *that it* is resistent to ticarcillin/clavulanate, ceftazidime, and piperacillin]

5. Association Rules and Adverse Drug Reactions

Adverse drug reactions (ADE) pose a serious problem for the health of the public and cause wasteful expenses (Wilson, Thabane & Holbrook, 2003). It is estimated that ADEs account for 5% of hospital admissions (Pirmohamed, Breckenridge, Kitteringham & Park, 1998), 28% of emergency department visits (Patel & Zed, 2002), and for 5% of hospital deaths (Juntti-Patinen & Neuvonen, 2002). In US only, ADEs result in losses of several billion dollars annually.

Due to their impact, ADE are monitored internationally in multiple sites. The Uppsala Monitoring Center in Sweden, a unit of the World Health Organization (WHO), mines data originating from individual case safety reports (ICSRs) and maintains *Vigibase*, a WHO case safety reporting database. Its activity started in 1978 and access to *Vigibase* is allowed for a fee.

At the {Food and Drug Administration} (FDA), a US federal unit, the *AERS* database (Adverse Event Reporting System) is maintained where access is free. Besides, proprietary ADE databases exists at various pharma entities who, by US law, must record adverse reactions to drugs.

We discuss the study performed in (Harpaz, Chase & Friedman, 2010) and the observations of (Wilson, Thabane & Holbrook, 2003) on using association rules for mining ADE databases.

ADE can involve single or multiple drugs and describe single or multiple adverse reactions. The simplest association rule describing an ADE has the form *Vioxx* → *heart attack* and involves one drug and one reactions. Clearly, rule of this form cannot capture ADE that result from undesirable drug interactions and this is the focus of (Harpaz, Chase & Friedman, 2010). This study is based on a set of 162,744 reports of suspected ADEs reported to AERS and published in the year 2008. A total of 1167 multi-item ADE associations were identified.

An ADE database has certain unique characteristics that allow for more efficient mining algorithms. Namely, the set of items is partitioned into two classes: *drugs* and *symptoms*; association rules have the form $D \rightarrow S$, where S is a set of D drugs and D is a set of symptoms.

Given a set of drugs it is important to find the largest set of symptoms such that $S \rightarrow D$ has a certain level of support and confidence. Indexing based on drugs and on symptoms was used to speed up searches in the data.

The general architecture of the AERS database is shown in Figure 1.

The attribute that binds various parts of the AERS database is: the unique number for identifying an EARS report.

A taxonomy that characterizes the associations was developed based on a representative sample, as shown in the next two tables.

Fig. 1: Architecture of AERS database

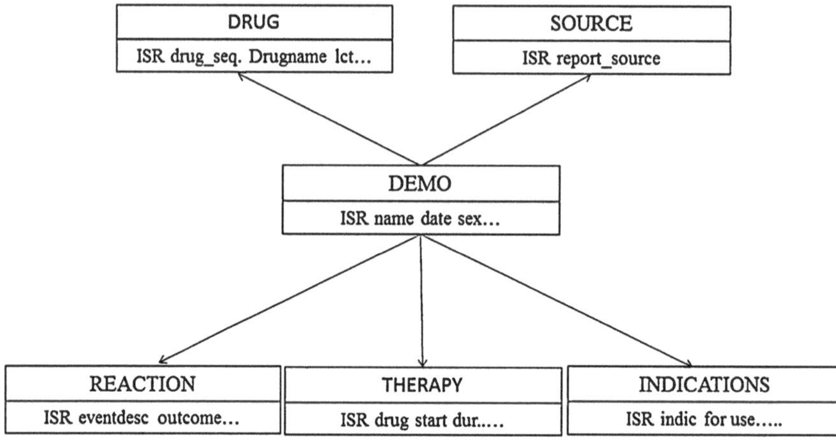

Taxonomy of multi-item sets of drugs

1a	Drug-drug interactions found that are known	4%
1b	Drug-drug combinations known to be given together or treat same indication	78%
1c	Drug-drug combinations that seem to be due to confounding	9%
1d	Drug-drug interactions that are unknown	9%

Taxonomy of multi-item ADE associations rules

2a	Associations (drug[s]-event) that are known	67%
2b	Associations (drug[s]-event) that are unknown	33%

67% of potential multi-item ADE associations identified were characterized and clinically validated by a domain expert as previously recognized ADE associations.

Filtering of the rules was done based on interestingness measures (confidence is just one of them). Actually, in this case confidence is inappropriate since rules of the form $X\ S \rightarrow$ *nnauseaausea* will have high confidence due to the high frequency of *nausenausea*.

Various alternatives for choosing an interest measure for association rules are studied extensively in data mining (Tan, Kumar & Srivastava, 2002; Bayardo &

Agrawal, 1999), (Jaroszewicz & Simovici, 2004; Hilderman & Hamilton, 1999; Jaroszewicz & Simovici, 2001).

Let $XX \rightarrow YY$ be an association rule. Denote the supports of the item sets $XX \cap YY$, $XX \cap \bar{Y}$, $\bar{X} \rightarrow Y$, $\bar{X} \rightarrow \bar{Y}$ by *a, b, c, d* respectively. The most commonly used interestingness measures for $XX \rightarrow YY$ are given next.

Interestingness Measure	**Formula**
Support	aa
Confidence	$\frac{a}{a+c}$
χ^2	$\frac{(aadd - bcbc)^2 (aa+bb+cc+dd)}{(aa+bb)(aa+cc)(bb+cc)(bb+dd)}$
interest (lift)	$\frac{aa(aa+bb+cc+dd)}{(aa+bb)^2}$
Conviction	$\frac{(aa+cc)(bb+dd)}{(aa+bb+cc+dd)ac}$

In Harpaz, Chase & Friedman (2010) the interestingness measure used was the *Relative Reporting Ratio* (RRR), defined by

$$RRRRRR = \frac{nn \cdot supp\ XY\ supp(XY)}{supsuppp(XX)\ supp\ s(YY)} = \frac{(aa+bb+cc+dd)(nn-dd)}{(aa+bb)(aa+cc)}$$

where $nn = aa + bb + cc + dd$ is the total number of records.

Note that RRR can be written as

$$RRRRRR = confconf(XX \rightarrow YY) \cdot \frac{nn}{suppsupp(YY)}$$

and can be regarded as the confidence of the rule $XX \rightarrow YY$ normalized by the relative support of the consequent. RRR is symmetric relative to X and Y.

Large values of RRR suggest that the occurrence of drugs-adverse reactions is larger than in the general collection of drugs. – A sample of multi-item ADE associations found in (Harpaz, Chase & Friedman, 2010) is:

1a–2a	metformin metoprolol → *nausea*	50	7.4
1b–2a	cyclophosphamide prednisone vincristine → *febrile neutropenia*	78	45
1c–2a	cyclophosphamide doxorubicin prednisone rituximab → *febrile neutropenia*	63	59
1b–2b	atorvastatin, lisinopril → *dyspnoea*	55	3.5
1a–2b	omeprazole simvastatin → *dispnoea*	58	12
1d–2b	varenicline darvocet → *abnormal dreams fatigue insomnia mem.impairment nausea*	52	2668

Since both *metformin* and *metoprolol* cause nausea, association rules of the form is metformin metoprolol → nausea is forseeable.

The rule

cyclophosphamide prednisone vincristine → febrile netropenia

involves a drug combination used in cancer treatment and describes a known complication. Similar conclusions are obtained for a variety of combination of other drugs.

6. Transitivity of Association Rules

A study of interactions between medications, laboratory results and problems using association rules was done by Wright, Chen, and Maloney at BWH in Boston (Wright, Chen & Maloney, 2010). The data examined included 100,000 patients. Encoding of problems, laboratory results, and medications was done using proprietary terminologies.

The importance of this study is that it highlighted difficulties of inferences involving probabilistic implications expressed by association rules. The authors noted that certain association rule occur with an unjustified high level of confidence. A typical example is the AR *insulin → hypertentsion* which involves unrelated terms. The explanation is the existence of co-morbidities, in this case, diabetes and hypertension, which highlights the need of mining for co-morbidities. It is shown that item sets such that

{lisinopril, multivitamin, hypertension}
{insulin, metformin, lisinopril, diabetes, hypertension}
{insulin, diabetes}
{metformin, diabetes}
{metformin, polycystic ovarian syndrome}

occur with a high level of support.

The difficulty of analyzing such association rules comes from the fact that association rules do not enjoy transitivity. This means that if $X \rightarrow Y$ and $Y \rightarrow Z$ are association rules with known confidence no conclusion can be drawn about the confidence of the AR $X \rightarrow Z$, as we show in the next example.

Example:
For the data set

	A	**B**	**C**
t_1	1	1	0
t_2	0	1	1

$A \rightarrow B$ and $B \rightarrow C$ we have $supp(A \rightarrow B) = 50\%$, $conf(A \rightarrow B) = 100\%$ and $supp(B \rightarrow C) = 100\%$, $conf(B \rightarrow C) = 50\%$, but $supp(A \rightarrow C) = 50\%$ and $conf(A \rightarrow C) = 0$.

On the other hand, for the data set

	A	**B**	**C**
t_1	1	0	1
t_2	0	1	0

and $A \rightarrow B$ and $B \rightarrow C$ we have $supp(A \rightarrow B) = 50\%$, $conf(A \rightarrow B) = 0$ and $supp(B \rightarrow C) = 50\%$, $conf(B \rightarrow C) = 0$, but $supp(A \rightarrow C) = 50\%$ and $conf(A \rightarrow C) = 100\%$. Thus, the confidence of $A \rightarrow C$ is unrelated to either $conf(A \rightarrow B)$ or to $conf(B \rightarrow C)$.

To deal with the lack of transitivity, it is necessary to investigate association rules of the form $X \rightarrow Z$ starting from existent AR $X \rightarrow Y$ and $Y \rightarrow Z$ which have a satisfactory medical interpretation. This is the point of view espoused in (Narayanasamy, Mukhopadhyay, Palakal & Mostafa, 2004) who present their software *TransMiner*.

The reverse approach is adopted in (Wright, Chen & Maloney, An automated technique for identifying associations between medications, laboratory results and problems, 2010): starting from an association rule $X \rightarrow Z$ (e.g. *insulin* → *hypertension* they seek to identify candidate item sets Y such that $X \rightarrow Y$ and $Y \rightarrow Z$ are plausible association rules. Y could be diabetes or other co-morbidities of hypertension; once these cases are excluded the confidence of *insulin* → *hypertension* decreases sharply.

7. Conclusions and Open Problems

DM cannot replace the human factor in medical research; however it can be a precious auxiliary instrument in epidemiology and pharmacovigilance. Interaction between DM and medical research is beneficial for both domains; biology and medicine suggest novel problems for data mining and machine learning.

Many open problems remain to be resolved. We estimate that extending association mining to unstructured data (progress reports, radiology reports, operative notes, outpatient notes), integration of "gold standards" in evaluation of AR extracted from medical practice, developing information-theoretical techniques for AR evaluation will attract the interest of both data miners and medical researchers because of their potential benefits in the practice of medicine. We conclude with a quotation from (L. L. Weed, vol. 315, 1997) written 29 years after the motto of this paper:

Knowledge should be held in tools that are kept up to date and used routinel – not in heads, which are expensive to load and faulty in the retention and processing of knowledge.

References

Adamo, J. M. (2001): Data Mining for Association Rules and Sequential Patterns. New York: Springer.

Agarwal, R. C.; Agarwal, C. C. & Prasad, V. (2001): A tree projection algorithm for generation of frequent item sets. In: Journal of Parallel and Distributed Computing, 61, pp. 350–371.

Agarwal, R. & Schaffer, J. (1996): Parallel mining of association rules. In: IEEE Transactions on Knowledge and Data Engineering, 8, pp. 962–969.

Agrawal, R.; T. Imielinski & Swami, A. N. (1993): Mining association rules between sets of items in large databases. In: Proceedings of the 1993 ACM SIGMOD International Conf. on Management of Data (pp. 207–216). Washington, D. C.: ACM.

Bayardo, R. & Agrawal, R. (1999): Mining the most interesting rules. In: Proceedings of the 5th KDD (pp. 145–153). San Diego: ACM.

Brosetts, S. E. & Hymel, P. A. (2008): Data Mining and Infection Control. In: Cliniscs in Laboratory Medicine, 28, pp. 119–126.

Brosetts, S. E.; Sprague, A. P.; Hardin, J. M.; Waites, K. B.; Jones, T. & Moser, S. A. (1998): Association Rules and data mining in hospital infection control and public health surveillance. In: Journal of American Medical Information Association, 5, pp. 373–361.

Han, J., Pei, J. & Yin, Y. (2000): Mining frequent patterns without candidate generation. In: W. Chen, J. F. Naughton & P. A. Bernstein (Ed.), Proceedings of ICDM (pp. 1–12). Dallas, TX: ACM.

Harpaz, R.; Chase, H. S. & Friedman, C. (2010): Mining multi-item drug adverse effect associations in spontaneous reporting systems. In: BMC Bioinformatics, 11.

Hilderman, R. & Hamilton, H. (1999): Knowledge discovery and interestingness measures: a survey. University of Regina, Computer Science.

Jaroszewicz, S. & Simovici, D. A. (2001): A general measure of rule interestingness. Principles of Data Mining and Knowledge Discovery, LNAI 2168 (pp. 253–266). Heidelberg: Springer.

Jaroszewicz, S. & Simovici, D. A. (2004): Interestingness of frequent itemsets unsing Bayesian networks as background knowledge. In: Proceedings of the 10th KDD (pp. 178–186). Seattle, WA: ACM.

Juntti-Patinen, L. & Neuvonen, P. J. (2002): Drug-related deaths in a university central hospital. In: European Journal of Clinical Pharmacology, 58, pp. 479–482.

L. L. Weed, M. (vol. 315, 1997): New connections between medical knowledge and patient care. In: British Medical Journal, pp. 231–235.

Narayanasamy, V.; Mukhopadhyay, S.; Palakal, M. & Mostafa, J. (2004): Trans-miner: Mining transitive association rules among biological objects from medline. In: Journal of Biomedical Science, 11, pp. 864–873.

Patel, P. & Zed, P. J. (2002): Drug-related visits to the emergency department: how big is the problem? In: Pharmacotherapy, 22, pp. 915–923.

Pirmohamed, M.; Breckenridge, A. M.; Kitteringham, N. & Park, B. K. (1998): Adverse Drug Reactions. In: British Medical Journal, 316, pp. 1295–1298.

Simovici, D. & Djeraba, C. (2008): Mathematical Tools for Data Mining. London: Springer.

Steinbach, M.; Karypis, G. & Kumar, V. (2000): A comparison of document clustering techniques. In: M. Grobelnik, D. Mladenic, & N. Milic-Freyling (Ed.), KDD Workshop on Text Mining. Boston.

Tan, P., Kumar, V. & Srivastava, J. (2002): Selecting the right interestingness measure for association patterns. In: Proceedings of KDD, (pp. 32–41).

Wilson, A. M.; Thabane, L. & Holbrook, A. (2003): Applications of data mining techniques in pharmacovigilance. In: British Journal of Clinical Pharmacology, 57, pp. 127–134.

Wright, A.; Chen, E. S. & Maloney, F. L. (2010): An automated technique for identifying associations between medications, laboratory results and problems. In: Journal of Biomedical Informatics, 43, pp. 891–901.

Wright, A.; Chen, E. S. & Maloney, F. L. (2010): An automated technique for identifying associations between medications, laboratory results and problems. In: Journal of Biomedical Informatics, 43, pp. 891–901.

Zaki, M. J. & Hsiao, C. J. (2005). Efficient algorithms for mining closed itemsets and their lattice structure. In: IEEE Transactions for Data and Knowledge Engineering, 17, pp. 462–478.

How to Promote Life Sciences as a Driving Force for Innovation

Three sectors – Infinite opportunities: Private-Government-Academic Opportunities to Advance the Life Sciences

Robyn Hannigan

1. Introduction

Major advances in basic life sciences occur at the intersections of disciplines. This requires not only collaboration across research disciplines but, increasing, demands collaboration across research sectors. As opportunities emerge within the life sciences to advance biological theory and our understanding of the natural world academia is increasingly partnering with the private and public sector. Given the cost of research and development within the private and public sector, partnerships with academic researchers enables investigation of "risky" research, affords nimble translation of intellectual property, and rapid infusion of new technologies into the marketplace. The capacities of academia as partners in basic research are further enhanced by their role in training the next generation of scientists thereby providing private and public sectors impact on the training and development of this workforce.

The role of academia in research and development is varied. In addition to developing human capital, academia must integrate with the private sector and more increasingly is involved in commercialization of technologies. More importantly academia is critical to the development of new technologies and new knowledge upon which the public and private sector depend. Despite the increasing role of academia in collaborative research with public and private partners few academic institutions provide training to its faculty regarding the opportunities and challenges of entrepreneurial science. Even more importantly most fail to provide such training to the next generation of researchers through focused innovation training opportunities for undergraduate and graduate science students. Engineering programs are built upon the translation of knowledge to action and so have a long history of providing such training. Science programs, however, fail to capitalize on these best practices and in so doing fail to prepare students for the realities of public/private research and development.

This chapter will discuss the role of entrepreneurial science in academic-private-public research and development partnerships and the critical aspects of training that must be provided to ensure that basic science research is translated to the development of technology and solutions to address our most pressing issues. Entrepreneurship and innovation are the foundation of economic prosperity and are also central to the academic enterprise. Though seldom couched in these terms the role of academic research is to be innovative, to pursue risky research agenda's that are focused on advancing theory within the discipline. Academics must, by definition, be entrepreneurial in funding their research programs. As private and public research and development move into the "new economy" which is driven by the search for opportunity rather than the avoidance of risk (Bygrave 1998) it is not unexpected that these sectors are reaching out to academia and that academic institutions are becoming more cognizant of the role of entrepreneurship in balancing their research portfolio.

2. Entrepreneurism and Academia

Over the coming years we expect to see a shift in research and development economies away from the traditional triumvirate of big business-government-unions to one that actively engages, and indeed relies upon, universities and university-based start-up companies (Figure 1). Whether driven by necessity top gain access to additional sources of funding and revenue or simply as a natural progression in research and development in today's global economy, Universities are increasingly expected to play a role in knowledge and technology development and transfer. In recent years a number of large established businesses have reached out to Universities, capitalizing on the capacity of academia to pursue high risk-high payoff research projects.

Fig. 1: Increasingly Universities are partners in not only government-funded research but industry-led research and development. In addition to start-up businesses evolving from academic research we are seeing increasingly more direct partnerships between the private and academic sectors, often funded through government contracts and grant programs [after (Schramm 2006)]

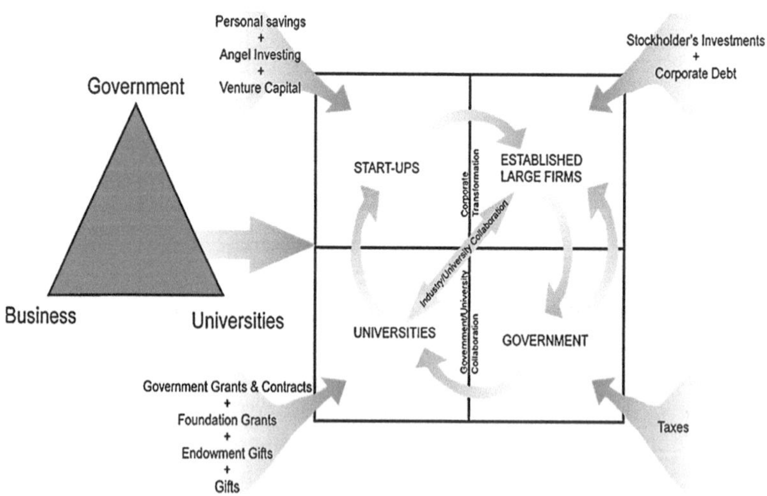

Over hundreds of years academic institutions have provided a constant flow of ideas for new businesses. In the mid-20th century, with the investment of federal and state government funds, academic researchers began to actively engage in research and development ensuring the translation of knowledge discovery to practical application. With the Bayh-Dole Act of 1980 the process of technology transfer was great accelerated with this acceleration accompanied by expansion of entrepreneurial activities within University settings. This cultural shift results in significant economic growth. Witness the fact that start-up companies spun-off from just one institution, Massachusetts Institute of Technology (MIT), would constitute a nation with the 24th largest GDP in the world (Schramm 2004).

While fostering entrepreneurial science, in an academic setting, is becoming increasingly important for all research Universities it is not a "panacea" for addressing the funding shortfalls that academic institutions currently face. Rather entrepreneurial science should be considered routine business for any academic institution that pretends to lead research and development in any science field. Providing students with training in entrepreneurial science, ensuring faculty and students are trained in responsible conduct of research including technology transfer and intellectual property, and providing the expertise and infrastructure required to foster the development of new knowledge and technologies with potential commercial value should be part of the "business-as-usual" practice of research institutions. These efforts should, however, not undermine the primary role of academic institutions as stewards of knowledge and should not undermine the academic freedom and independence that is crucial for research and development. It is the independence and rigor of academic research institutions that specifically attract private and public sector partners and the ethics of academic research should never be compromised.

3. Universities and Technology Transfer

Since 1969 U.S. Universities and Colleges have been awarded 4 million utility patents (issued for invention of a new or useful process, machine, manufacture, or composition of matter, or a new and useful improvement thereof). Though this number seems large it represents only 1.5% of the total US Utility Patents (USPTO 2012). Between 1969 (pre-Bayh-Dole) and 1994 U.S. academic institutions were granted 17183 patents (0.88% of the total). Since 1994 this number has risen to a maximum of 2.2% of the total. Despite the fact that Universities, since the Bayh-Dole Act, have sought to commercialize virtually any intellectual property developed by their faculty the success rate remains very low. The intent is that the University, once securing the patent, can license the technology to industry through exclusive or non-exclusive licensing agreements and receive royalties or other payment in return. While considered, initially, as a win-win for Universities and a potential solution to declining research funding most Universities have failed to recover the costs associated with maintaining active technology transfer offices as well as the costs of filing patent applications. Most licenses patents will not result in financial returns to the University and the vast majority of University-owned patents will never be licensed by industry at all.

In fiscal year 2010 the Association of University Technology Managers (AUTM 2010) found university licensing and startup activity to be quite strong with the number of new start-up companies, licenses, and options above those in the previous year (180 member Universities). This increase was also accompanied by an increase in research expenditures and the highest increase in U.S. University owned patents awarded since 1993. Taking data from 2009 and 2010 the success rate of converting a patent disclosure to a patent application is 50% with only a 38% success rate. Considering that an average of 312 million dollars in legal fees are expended to secure patents this may seem a poor use of University funds. However, across all surveyed members AUTM found that royalty payments averaged 1.5 billion dollars across 2009–2010. The potential revenue provided by intellectual property is all too often the reason Universities support active technology transfer offices. It is important to note, however, that only 0.6% of the licenses generated more than $1 million in revenue. Technology transfer costs money and the potential losses are high. For example, Stanford University expends, on average, $2.6 million per year to operate its technology transfer office and spends an additional $5 million on patenting costs. Stanford did earn significant licensing revenue ($41.2 million in FY2000) but is an exception, not the rule. How many Universities could maintain expenditures of 8–10 million dollars annually in return for licensing revenues less than $1 million annually? Few if any could and, interestingly, few if ever actually bear the cost alone. More than 70% of patents filed by industry, in the US, are based on discoveries made by University researchers funded by federal grants (Narin et al. 1997). In Europe companies attribute less than 5% of their innovation to Universities (EUROSTAT 2004).

Given that technology transfer, at most academic institutions is a "loss leader", there are opportunities presented by fostering entrepreneurship among students and faculty that provide significant, albeit non-monetary, benefits. First is that investing in high-risk/high-payoff research by faculty may translate to monetary return. More often, however, such research expenditures garner reputation building capital that the University can leverage to increase endowments and advance research and development activities through industry-government-academic partnerships. Perhaps more important are the benefits to students who are trained in both science and business through the experiences gained in working in an entrepreneurial environment. These students gain critical work skills and have a deeper understanding of the soft-skills required to succeed upon graduation. Entrepreneurial science teaches them team work, project management, reporting, financial management, and intellectual property skills that directly translates to workforce needs within an increasingly technology driven economy. Producing high quality scientists is the primary role of academic institutions and fostering entrepreneurship provides additional reputation building capital upon which the University can build its capital campaigns.

4. Open Innovation and Cross-Sectoral Partnerships

As our global economy becomes increasingly dependent on innovation we see the old system of linear innovation (research and development controlled by a single company) replaced by "open innovation" (Chesbrough 2003). In industry the competitive edge goes to the most innovative business models and as such open innovation demands collaboration between Universities and industry. Patents are used, today, not to protect market share but to facilitate the trade of intellectual property. This shift in perspective, for industry, is a motivating factor behind their desire to collaborate with Universities and their willingness to bear much of the cost of intellectual property development and technology transfer. In addition to collaborating on high risk, potentially transformative research University spin-offs represent, to industry, useful opportunities to test new business models and, particularly when students are engaged, to rejuvenate the economy (Capart 2006).

The filing of patent applications by universities is not incompatible with their mission to produce and disseminate new knowledge provided that universities are ethical in their intellectual property policies and that cross-sectoral partnerships are mutually beneficial. Universities should only consider filing a patent application if the filing increases the probability that applications of the technology will be developed. They should not, and are not prepared to, speculate on market value. Universities, once an application is filed, must actively seek partners to develop the inventions and grant licensing terms that are compatible with the public interest. This is particularly important for public U.S. Universities who are a public trust. Universities should only grant exclusive licensing rights on a limited bases and only in exchange for specific, ethical commitments from the licensing partner. University inventors must be engaged in the commercialization process directly or receive equitable revenue through the licensing agreement to motivate and encourage innovation and knowledge transfer. These funds should never be used, however, as a substitute for external, peer-reviewed research funding by University researchers. Lastly, Universities must, in their technology transfer office, employ and engage trained professionals who deeply understand business development across relevant sectors. In the case of patent filings quality is far more important than quantity.

Cross-sectoral partnerships are critical to Universities participation in open innovation and in the 21st century economy. Whether through training of scientists, engaging faculty in intellectual property development, or advancing technology development through industry collaborations (Figure 2), academia provides vital skills and opportunities to both government and industry research and development programs.

Fig. 2: Knowledge creation and maturation requires not only academic expertise and investments through intellectual capital development but by providing industry and government partners opportunities to leverage market needs into advancing both workforce development as well as rapid innovation and product development. Adapted from (Seline 2009)

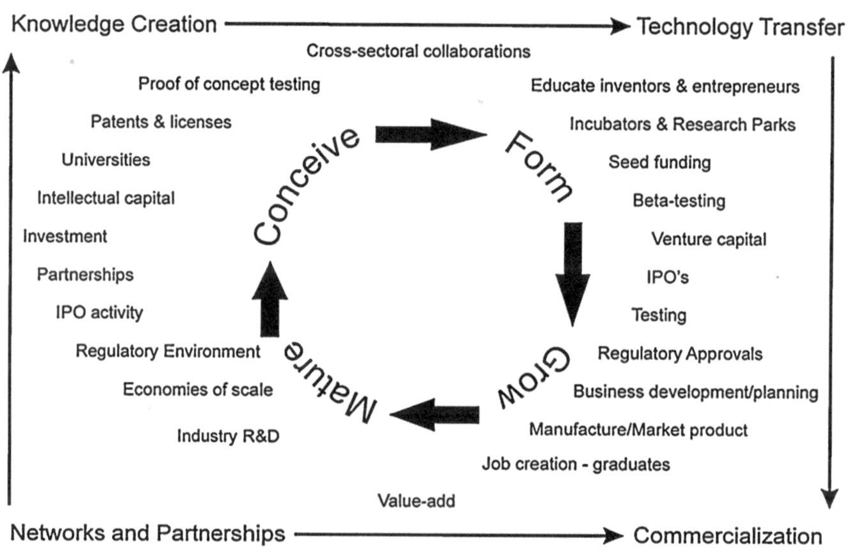

5. Summary and Recommendations

There can be no doubt that University technology transfer and commercialization activities can positively impact local, state, and national economies. Whether through revenue-generating partnerships with big business (e.g., Stanford

University - Google, Inc.), start-up companies spun off from Universities (e.g., MIT), or through workforce development, academic institutions play an increasingly important role in innovation, particularly in the life sciences. If Universities intend to capitalize on the culture of open innovation and play a role in the knowledge-based economy they must have a strong a focused University research base that directly feeds into commercialization. Universities should not try to be all things but should target areas which they currently have the expertise to build upon and then target hires, federal funding programs, industry-sponsored research collaborations, and state-based initiatives to solidify their position as innovation leaders in a particular area. Since federal funding, in the U.S., forms the base for research and development at Universities these institutions must be fully aware of on-going and planned changes to agency funding priorities. Research universities must play a strong role in economic development and be able to point to a champion for the university's role in the regional economy. This champion should engage local communities as well as the public and private sector organizations who benefit from university research and development to foster a cultural of cross-sectoral partnerships for the public good. Universities should directly invest in entrepreneurial faculty and students through both capital and attention being sure to showcase their research to potential investors. Universities should extend the business school model of "executives in residence" to include entrepreneurial researchers in residence and structured entrepreneurial faculty positions. For a university to be successful in technology transfer the culture of entrepreneurship must be fostered both within the faculty as well as the students and must be accompanied by implicit and explicit rewards and incentives for faculty who participate in technology transfer and commercialization. The University must provide networking opportunities for entrepreneurial faculty and students top meet with investors, potential clients, and other entrepreneurs. Lastly, and most importantly universities must have strong entrepreneurship programs that extend beyond the business school to include and actively engage science and engineering students. If a university cannot accommodate the above recommendations then they should enter into the technology business with caution. Without the faculty research expertise, without technology transfer know-how, with no academic programming or incentive programs, and without a high-level champion these efforts are doomed to failure. By strategically aligning expertise and economic need with infrastructure and university culture, cross-sectoral partnerships and entrepreneurial science will lead Universities to advancing scientific discovery, serving the public good, and ensuring that our next generation of scientists is nimble, innovative, and collaborative.

References

AUTM, F. (2010): 2010 licensing activity survey. Retrieved July 31, 2012.

Bygrave, B. (1998): Stockton lecture. Business Strategy Review 9: pp. 11–18.

Capart, G. (2006): Should universities file patent applications? Ethical Perspectives: Journal of the European Ethics Network 13: pp. 221–230.

Chesbrough, H. W. (2003): Open innovation - the new imperative for creating and profiting from technology. Cambridge, MA, Harvard Business School Publishing Corporation.

EUROSTAT (2004): Innovation in europe: Results for the eu, iceland, and norway (1998–2001). Luxenbourg, Office for Official Publications of the European Communities: p. 298.

Narin, F., Hamilton, K. S., Olivastro, D. (1997): The increasing linkage between u.S. Technology and public science. Research Policy 26: pp. 317–330.

Schramm, C. J. (2004): Building entrepeneurial economics. Foreign Affairs 83: pp. 104–115.

Schramm, C. J. (2006): Entrepeneurial capitalism and the end of bureaucracy: Reforming the mutal dialog of risk aversion. American Economic Association. Boston, MA.

Seline, R. (2009): From the stimulus economy to the knowledge economy: Why tecna is even more critical. Technology Leadership. Washington, DC, Technology Councils of North America.

USPTO (2012): U.S. Colleges and universities - utility patent grants 1969–2008. Retrieved July 31, 2012.

Privatization of Universities and a New Social Compact: Trends in Funding

Robert F. Rich

There is no good news concerning the prospects for higher education funding for public universities in the United States. This paper focuses on the response of some institutions of public higher education in the United States given the trend toward decreasing support from state general revenue funds (GRF): increased reliance on revenue generated thru tuition especially from out-of-state and international students, increased reliance on grants and contracts as well as endowments, and increased autonomy as evidenced thru deregulation. Some experts have characterized these trends as a move toward privatization.

1. Privatization and Higher Education

The literature suggests that privatization has been an issue for over a decade and that there is no neat and clean way to characterize the success or failure. In some studies, the argument is made that one should use the term "autonomy" instead of privatization.

By "privatization" of public university, one or more of the following dimensions are most often referred to:

- Relying primarily on revenues generated by "tuition" as compared with general revenue from state government;
- Attracting a large percentage of "out-of-state" and "international" students who will pay full tuition;
- Attracting large amounts of research grants and contracts that can, in turn, generated ICR (indirect cost recovery or overhead); and
- Increasing the endowment of the university which, in turn, will generate income

In essence, this is a new model for the organization and financing of public universities.

2. Overview: State Support of Public Universities

As we think about this option, we should look at how state support of higher education has changed over time in publicly supported universities: The chart below is from the *Chronicle of Higher Education*. This chart presents percentages of state support of each institution's total budget:

	State Support		
	2001–2002	**2007–2008**	**% Change**
University of Washington	19%	16%	-15.79%
University of California at Berkeley	42%	30%	-28.57%
University of Illinois at Urbana-Champaign	30%	19%	-36.67%
University of Michigan	17%	12%	-29.41%
University of Texas at Austin	24%	17%	-29.17%
University of Virginia	19%	12%	-36.84%
University of Wisconsin at Madison	25%	20%	-20.00%
Research Level Universities	36%	32%	
Masters Level Universities	44%	39%	

This table indicates that major public universities have suffered major declines in state appropriations as a percentage of total university budgets, based on figures from the *Chronicle of Higher Education*. As of 2007–2008, state appropriations represented from 12 percent of the total university budget at the University of Michigan to 30 percent at the University of California – Berkeley. All systems listed in the table saw erosion on this indicator, from -16 percent to -37 percent between 2001–2002 and 2007–2008.

Using figures from *Grapevine*, we can see a bit different picture, in the table below, if we look at state appropriations to each of the above institutions over the same time period.

	University	Comm Coll	Total
Arizona State University			
2001–2002	416,765	135,578	949,926
Percent of Total Higher Education Appropriation	43.87%	14.27%	
2007–2008	416,765	149,991	1,267,786
Percent of Total Higher Education Appropriation	32.87%	11.83%	

	University	Comm Coll	Total
Oregon State University			
2001–2002	359,094	232,519	714,837
Percent of Total Higher Education Appropriation	50.23%	32.53%	
2007–2008	338,884	253,988	720,365
Percent of Total Higher Education Appropriation	47.04%	35.26%	
University of Arizona			
2001–2002	287,998	135,578	949,926
Percent of Total Higher Education Appropriation	30.32%	14.27%	
2007–2008	362,389	146,991	1,267,786
Percent of Total Higher Education Appropriation	28.58%	11.59%	
University of California at Berkeley			
2001–2002	549888	2849302	9468062
Percent of Total Higher Education Appropriation	5.81%	30.09%	
2007–2008	474317	3916584	11062155
Percent of Total Higher Education Appropriation	4.29%	35.41%	
University of Illinois at Urbana-Champaign			
2001–2002	399508	406358	2922599
Percent of Total Higher Education Appropriation	13.67%	13.90%	
2007–2008	355651	406788	2935504
Percent of Total Higher Education Appropriation	12.12%	13.86%	
University of Michigan			
2001–2002	415624	319596	2273532
Percent of Total Higher Education Appropriation	18.28%	14.06%	
2007–2008	323440	293169	2043409
Percent of Total Higher Education Appropriation	15.83%	14.35%	
	University	**Comm Coll**	**Total**
University of Texas at Austin			
2001–2002	300456	967962	5074633
Percent of Total Higher Education Appropriation	5.92%	19.07%	
2007–2008	353881	1202474	6058375
Percent of Total Higher Education Appropriation	5.84%	19.85%	

	University	Comm Coll	Total
University of Virginia			
2001–2002	160874	314107	1681646
Percent of Total Higher Education Appropriation	9.57%	18.68%	
2007–2008	144176	375630	1870633
Percent of Total Higher Education Appropriation	7.71%	20.08%	
University of Washington			
2001–2002	326646	516649	1373895
Percent of Total Higher Education Appropriation	23.78%	37.60%	
2007–2008	351769	707807	1787124
Percent of Total Higher Education Appropriation	19.68%	39.61%	

This table shows that actual dollar declines in state appropriations have been less severe than as the percentage of total budgets, as reported in *Grapevine*. Indeed, several campuses saw gains in appropriations in the 2001–2002 to 2007–2008 period, as at Arizona, Texas-Austin and Washington. On the other hand, Berkeley saw an erosion of 13.7 percent in its state appropriation in the same period while Michigan saw its state appropriation drop by 22 percent.

During the 2001–2008 period, community college systems in all the states listed saw their appropriations grow with the exception of Michigan. California and Washington saw robust increases of 37 percent in funding for their community colleges while Texas increased support for its community colleges by 24 percent.

3. Tuition Considerations

Tuition revenue has played an important role in offsetting much of the loss of state support for many universities. As we face a period where additional declines in state GRF are likely, will it be possible for tuition revenue to play a similar role in offsetting those reductions? There are several ways to assess the capacity of tuition income to grow at rapid enough rates to offset GRF declines. First, how does Illinois tuition compare with our public peers? Second, to what extent do we have the capacity to attract more out-of-state and international students who might pay full tuition? Finally, does our growing cost put at risk our capacity to fill our classes?

When comparing our costs to those of other public Big Ten institutions, Illinois now has the second highest total cost of attendance. However, many students at Illinois pay tuition differentials based on their majors. The differentials for Engineering, Business and Sciences are the highest in the Big Ten. When considering those differentials, the University of Illinois at Urbana-Champaign (UIUC) tuition is the highest. A comparison with all Association of American Universities (AAU) publics would have a similar outcome, placing Illinois third until differentials are taken into account.

Cost of Attendance for In-State Freshmen 2009–2010

	Tuition/Fees	Room & Board	Books	Other	Total	Rank
Illinois	$12,660 ($16,978)	$9,284	$1,200	$2,510	$25,654 ($29,972)	2 (1)
Indiana	$8,614	$7,580	$792	$3,148	$20,134	9
Iowa	$6,824	$8,004	$1,090	$3,520	$19,438	11
Michigan	$11,659	$8,924	$1,048	$2,090	$23,721	4
Michigan State	$11,434	$7,444	$962	$1,708	$21,548	6
Minnesota	$11,466	$7,392	$1,000	$2,194	$22,052	5
Nebraska	$7,312	$8,196	$1,010	$3,368	$19,886	10
Ohio State	$8,658	$9,468	$1,440	$4,572	$24,138	3
Penn State	$14,416	$8,820	$1,360	$3,924	$28,520	1
Purdue	$8,638	$8,710	$1,220	$2,070	$20,638	7
Wisconsin	$8,313	$8,040	$1,040	$2,890	$20,283	8

Source: Office of Student Financial Aid

4. Enrollment and Tuition

As part of the movement toward autonomy, universities have been moving toward being able to set their own tuition. Here is a profile of how some of the public universities have set their tuition;

Tuition Setting	Decision Body
Arizona State University	Arizona Board of Regents
Oregon State University	Oregon University System
University of Arizona	Arizona Board of Regents
University of California at Berkeley	Regents
University of Illinois at Urbana-Champaign	Trustees
University of Michigan	Board of Regents
University of Texas at Austin	Tuition Policy Advisory Committee
University of Virginia	Board of Visitors
University of Washington	State
University of Wisconsin at Madison	UW Regents

As we know from the experience in Illinois, the matter is complicated by the guaranteed tuition program. We did not find any mention of other profiled universities having such programs limiting their ability to set tuition.

5. The Role of Out-of-State and International Students

Another important part of the equation, in terms of achieving more flexibility and autonomy, is the ratio of out-of-state and international students, who pay a much higher tuition rate, at each institution.

Out-of-State Students (there is no one single source with these numbers for all of these institutions):

Arizona State University	23%*
Oregon State	15%
University of Arizona	24%*
University of California at Berkeley	7%
University of Illinois at Urbana-Champaign	8%
University of Michigan	33%
University of Texas at Austin	15%
University of Virginia	28%
University of Washington	12%
University of Wisconsin at Madison	21%**

* *Western Undergraduate Exchange program instituted in 2009 for 14 western states to pay a discounted tuition rate for these out-of-state students*

** *Minnesota students counted as in-state*

Some recent developments with out-of-state student enrollments and policy changes include the following:

- The University of California at Berkeley is planning to admit more out-of-state residents this year. Berkeley's non-resident undergraduate population is quite low (around 10 percent typically, counting both U.S. residents outside California and international students). Robert Birgeneau, the chancellor, told The Contra Costa Times that he hopes that the shift will set off some anger from California residents, saying: "Actually, I hope for some pushback. This is connected to the state's failure to pay for the University of California."

- While faculty leaders at the University of California have been critical of the administration's management of the state's severe budget crisis, they too are calling for more out-of-state students. Department chairs at UC San Diego issued a statement calling for significant increases in out-of-state enrollments. "Admitting more students from other states would also enhance UCSD's national reputation and it will benefit California in the long run since a significant number of American students who go to university out-of-state end up settling in the state where they attended university. It is in California's interest to attract some of the best and the brightest high school graduates from araound the country, to provide them with a world-class education and then to reap the tax revenues that result when these university graduates enter California's workforce", said the statement.
- The University of Massachusetts at Amherst is planning to increase undergraduate enrollment by 15 percent over the next decade, with the additional students coming from out-of-state. If the plan succeeds, the share of undergraduate enrollment from outside Massachusetts would grow from 20 percent to 30 percent.
- Rutgers University, where out-of-state enrollment is currently about 10 percent, is planning a gradual increase to about 25 percent.
- The president of the University of Colorado wants the state to lift caps on non-resident enrollment.
- The comptroller of New York State issued a report last month calling on the State University of New York to increase non-resident tuition, with the goal of increasing revenues.

International Students (there is no one single source with these numbers for all of these institutions):

Arizona State University	2.45%
Oregon State	1.85%
University of Arizona	3.23%
University of California at Berkeley	5.49%
University of Illinois at Urbana-Champaign	13.00%
University of Michigan	5.70%
University of Texas at Austin	4.13%
University of Virginia	5.84%
University of Washington	5.35%
University of Wisconsin at Madison	8.00%

6. The Role of Financial Aid

As rates of tuition increase and as universities seek to attract more out-of-state and international students, the question of affordability arises. Accompanying this trend may be a much greater need for financial aid.

Given this concern, we looked at the percentage of students receiving financial support.

Percentage of Students Receiving Financial Aid	
Arizona State University	73%
Oregon State	67%
University of Arizona	80%
University of California at Berkeley	66%
University of Illinois at Urbana-Champaign	69%
University of Michigan	79%
University of Texas at Austin	67%
University of Virginia	55%
University of Washington	57%
University of Wisconsin at Madison	69%
Harvard	81%
Princeton	58%

7. The Role of University Endowments

As we consider “autonomy,” we should also look at the size of the endowment of other public institutions. Public universities may choose to use their endowment to off set costs in the university instead of tuition dollars and state appropriations.

The size of the endowment varies greatly from one university to the next:

Endowment	
Arizona State University	478 million
Oregon State University	455 million
University of Arizona	532 million
University of California at Berkeley	6.2 billion - U of C system
University of Illinois at Urbana-Champaign	1.45 billion - U of I system
University of Michigan	7.57 billion
University of Texas at Austin	16.1 billion -U of T System

Endowment	
University of Virginia	4.57 billion
University of Washington	2.2 billion
University of Wisconsin at Madison	1.73 billion - U of W Foundation

As we know from Illinois, the ability to use those funds for operating expenses in lean times is not a standard practice among most state university foundations (but is a practice among private universities).

8. Grants and Contracts

In any review of the revenues of universities, attention must be given to the area of grants and contacts. Not only do grants and contracts get frequent press mention for the institution that is obtaining a grant, the size of the grant is also likely to attract attention and possibly suggest to some that this might be an area where universities might be able to obtain more financial independence and autonomy.

The following chart shows the revenue from grants and contracts and the percentage increase in revenue from grants and contracts revenue from FY 02 to FY 08 for each institution in the cohort group for this report.

University	FY 02	FY 08	% Change
Arizona State University	$118,580,000	$ 267,442,000	125.54%
Oregon State University	$152,665,859	$ 183,331,183	20.09%
The University of Texas at Austin	$356,623,673	$ 527,548,132	47.93%
University of Arizona	$330,083,000	$ 446,519,000	35.27%
University of California at Berkeley	$402,738,000	$ 543,160,000	27.21%
University of Illinois at Urbana-Champaign	$377,807,759	$ 480,618,222	42.38%
University of Michigan at Ann Arbor	$675,366,047	$ 961,598,000	32.04%
University of Virginia – Main Campus	$260,211,721	$ 343,588,042	45.31%
University of Washington – Seattle Campus	$718,351,274	$1,043,848,734	45.27%

These percentage increases show that over a six year period, each institution was garnering anywhere from a 27% to an almost unbelievable 125% increase in revenues.

However, during this same period, the overall budgets of these institutions were also increasing at a similar pace.

The following chart shows the percentage of the total revenue for each university in the cohort that came from grants and contracts in FY 2002 and FY 2008:

University	FY 02	FY 08
Arizona State University	16.38%	18.70%
Oregon State University	30.91%	28.24%
The University of Texas at Austin	33.69%	31.76%
University of Arizona	33.75%	31.51%
University of California at Berkeley	30.68%	29.94%
University of Illinois at Urbana-Champaign	27.70%	25.97%
University of Michigan at Ann Arbor	19.08%	18.97%
University of Virginia – Main Campus	18.91%	15.22%
University of Washington – Seattle Campus	32.01%	32.54%
University of Wisconsin at Madison	29.72%	31.25%

These two charts illustrate that while there have been major increases in the revenue from grants and contracts, the actual percentage of the total revenue from grants and contacts for each university has remained fairly consistent and is likely not to have much bearing on changing revenue streams to obtain greater autonomy.

9. Mini-Case Studies from Several States

We have examined initiatives that have been undertaken in other states and think that highlighting some of these initiatives may be useful:

1. The state of Oregon reports that the 2010 appropriation for higher education is a "protected" 60 million dollars. Twenty years ago the appropriation was $63.3 million. This year's appropriation is just 34.9 million in 1990 dollars or a loss in real buying power of 42 percent. In response during the Spring of 2010, a white paper was prepared by the Oregon State Higher Education System proposing that a public endowment be created to help finance institutions of higher education. This white paper proposes a plan where the state funding commitment of $63 million a year can be used over thirty years to make annual debt payments of $800 million in general obligation bonds. The University of Oregon would match the 800 million in bond proceeds with money raised from private donors and manage the combined 1.6 billion as an endowment. The idea is that the endowment would generate revenue for the University which would be equal or

nearly equal to the amount that was being provided in state funding. The theory is that this "public endowment" would grow (an assumption of 9 percent for yield and 4 percent for distribution rate) to 6.9 billion over thirty years with an annual distribution of $263.4 billion dollars. This would, in turn, significantly increase the budget of the University of Oregon.

2. The Arizona Board of Regents is exploring how it can reshape higher education funding. Two ideas being explored, as reported at a Regents meeting on September 25, 2010, are a dramatic expansion of on-line education and a new taxing source for universities to add to or replace legislative funding. In addition, other ideas that are being explored include privatizing some of the university's professional schools, eliminating "under-performing" or duplicative academic programs, and doubling research grant dollars. More specifically, the Regents are considering operating the Arizona State University's Sandra Day O'Connor College of Law with no public investment. This would be accomplished by raising tuition and increasing enrollment. This would free up $13 million in annual state funding which could be used by other parts of the university. In addition, the Regents plan calls for all public universities to increase their on-line enrollment with Arizona State University adding 27,000 students to its on-line enrollment which would be up from the 3,000 currently enrolled. Finally, the university is proposing a new funding model tied to the university's performance such as increasing the share of bachelor's degrees. Funding could come as a legislative appropriation or it could come as a voter-approved tax. Arizona is at the front-end of this discussion so the plans are not very well worked out.
3. The State of Oklahoma offers state matching funds through the State Regents for Higher Education to privately endowed chairs and professorship donations. A recent Oklahoma Supreme Court ruling cleared the way for issuance of 50 million dollars in revenue bonds to provide funds for many endowed positions that had been awaiting a state match. Oklahoma University now had 403 endowed faculty positions up from 101 in 1994.

10. Other Dimensions to Consider

As we think about privatization or "autonomy," there are several other dimensions that should be considered:

- A trade-off between less reliance on state general revenue funds and "deregulation" or providing a university with more "flexibility," Colorado public universities have been provided with this type of flexibility in terms of: (a)

being able to raise tuition up to 9% in a given year without approval of the General Assembly; (b) Increasing the number of international and out-of-state students; and (c) an. "expedited process" for approval of university spending.
- Granting constitutional autonomy so that a university can set its own tuition rates (e.g., University of Michigan).
- Making up appropriations from the state with endowment income could be an almost impossible task. For example, the University of Michigan currently only receives 6% or 320 million dollars in its overall budget. However, in order to replace these funds, the University would need to generate additional endowment in order to increase its endowment to the level of 16 billion dollars. The current endowment is about 7.5 billion dollars.
- The University of Michigan attracts between 37 and 40% of out-of-state and international students at the undergraduate level. Who's paying full tuition? If one adds out-of-state students, nearly 50% of all students at the University of Michigan come from out-of-state.

Overall, there seems to be no clear trend that might be held up as a model for purposes of privatization.

11. Discussion

Public universities need to develop creative funding strategies which will allow them to maintain high levels of quality and continue to be incubators of innovation and economic development.

This will mean forging a new social compact with state legislatures and Governors (e.g., Maryland). The new social compact will call for a set level of "protected" state appropriation combined with some level of deregulation.

As Universities enter into new partnerships with state governments, other dimensions which will need to be considered will include: (a) increased numbers of out-of-state students and international students; (b) increased levels of grants and contracts which the University is able to attract; (c) increased levels of endowment which are generated from an aggressive fund-raising strategy; and (d) programmatic cuts and consolidations within the University.

References

Alexander, F. K. (2000): The changing face of accountability: Monitoring and assessing institutional performance in higher education. *Journal of Higher Education, 71*(4), pp. 411–431.

Banta, T. W., Rudolph, L. B., Van Dyke, J., & Fisher, H. S. (1996): Performance funding comes of age in Tennessee. *Journal of Higher Education,* 67(1), pp. 23–45.

Bogue, E. G. (2002): Twenty years of performance funding in Tennessee: A case study of policy intent and effectiveness. In: J. C. Burke (ed.), *Funding public colleges and universities: Popularity, problems, and prospects* (pp. 85–105). Albany, NY: SUNY Press.

Bogue, E. G. & Brown, W. (1982): Performance incentives for state colleges. *Harvard Business Review, 60*(6), pp. 123–128.

Bogue, E. G. & Johnson, B. D. (2009): *Performance incentives and public college accountability: A quarter century policy audit.* Presentation at the 2009 annual meeting of the Association for the Study of Higher Education, Vancouver, Canada.

Bogue, E. G., & Troutt, W. E. (1977): *Allocation of state funds on a performance criterion: Acting on the possible while awaiting perfection.* Paper presented at the Association for Institutional Research Forum.

Burke, J. C. (2002a): *Funding public colleges and universities: Popularity, problems, and prospects.* Albany, NY: SUNY Press.

Burke, J. C. (2002b): Performance funding in South Carolina: From fringe to mainstream. In: J. C. Burke (ed.), *Funding public colleges and universities: Popularity, problems, and prospects* (p. 195–219). Albany, NY: SUNY Press.

Dougherty, K. J., & Natow, R. S. (2010): *Continuity and change in long-lasting state performance funding systems for higher education: The cases of Tennessee and Florida* (CCRC Working Paper No. 18). New York, NY: Columbia University, Teachers College, Community College Research Center. Retrieved from http://ccrc.tc.columbia.edu/Publication.asp?UID=743.

Dougherty, K. J., Natow, R. S., Hare, R. J., & Vega, B. E. (2010): *The political origins of state-level performance funding for higher education: The cases of Florida, Illinois, Missouri, South Carolina, Tennessee, and Washington* (CCRC Working Paper No. 22). New York, NY: Columbia University, Teachers College, Community College Research Center.

Dougherty, K. J., Natow, R. S., Hare, R. J., Jones, S. M., & Vega, B. E. (2011): *The Politics of Performance Funding in Eight States: Origins, Demise, and Change.* New York, NY: Columbia University, Teachers College, Community College Research Center.

Harnisch, T. L. (2011): Performance-Based Funding: A Re-Emerging Strategy in Public Higher Education Financing. American Association of State Colleges and Universities, Policy Matters.

Hauptman, A. M. (2005): Performance-Based Funding in Higher Education. (Presentation).

McLendon, M.K., Hearn, J.C., & Deaton, R. (2006): Called to account: Analyzing the origins and spread of state performance accountability policies for higher education. *Educational Evaluation and Policy Analysis,* 28(1), pp. 1–24.

Midwestern Higher Education Compact (2009): Completion-Based Funding for Higher Education. Minneapolis, MN: Midwestern Higher Education Compact.

Rhoades, G. & Sporn, B. (2002): Quality assurance in Europe and the U.S.: Professional and political economic framing of higher education policy. *Higher Education, 43*(3), pp. 355–390.

Ruppert, S. (1995): Roots and realities of state-level performance indicator systems. *New Directions for Higher Education, 91,* pp. 11–23.

Serban, A. M. (1997): *Performance funding for higher education: A comparative analysis* (Doctoral dissertation), State University of New York at Albany. Available from ProQuest Dissertations and These database (UMI No. 9809292).

Serban, A. M. (1998): Performance funding criteria, levels, and methods. *New Directions for Institutional Research,* 1998(97), p. 61.

Simoneau, S. & Stevens, J. (2010): "Higher Education Governance Structure." Hartford, CT: Connecticut General Assembly.

South Carolina General Assembly (1996): Act 359, 111 Sess., R368 Cong. Rec.

South Carolina Legislative Audit Council (2003): *Follow-up: A review of the higher education performance funding process (June 2001)*: Columbia, SC: Author.

Zumeta, W. (2001): Public policy and accountability in higher education: Lessons from the past and present for the new millennium. In: D. Heller (ed.), *The states and public higher education policy: Affordability, access and accountability.* Baltimore, MD: Johns Hopkins University Press, pp. 155–197.

Financing Higher Education in Germany – Challenges and Perspectives

Hans Adam/Peter Mayer

1. Introduction

The quality of the educational system plays a key role for the competitiveness of nations. With structural changes taking place in the world economy and with comparative advantages of nations shifting rapidly, higher education systems become increasingly important for ensuring the success of nations aspiring to be economically successful and at the forefront of innovation. It is quite natural that almost all countries around the world discuss the effectiveness and efficiency of their higher education systems. One of the prominent themes within this debate is the issue of funding higher education.

In the past, the German higher education system has been recognized as one of the rather successful systems in the developed world, producing teaching and research output at a high and internationally recognized level. The system has produced a large number of highly qualified graduates and scholars, the research output has been considerable, and Germany's scholars have been well integrated into the international flow of knowledge. However, tertiary graduation rate is lower than in most other OECD countries, the number of students not completing their studies is considerable, students spend a high average number of years at university before they eventually graduate. And indicators such as the performance of German higher education institutions in internationally recognized rankings or the number of Germans receiving important prizes are sometimes interpreted as signs of weak performance of German higher education institutions.

This article addresses the issue of financing higher education in Germany. It looks at the total spending on tertiary education, and the various income sources for higher education institutions. Thereafter, economic theory is used to reflect on some of the key questions concerning the future of financing German higher education institutions.

2. The overall trend of financing higher education in Germany

The total allocation of funds for higher education of a nation is typically measured by using the ratio of total expenditure for higher education, i.e. spending irrespective of the sources of funds, to GDP (Figure 1). In 2008, Germany spent 1.2% of its GDP on tertiary education. With this figure, Germany was well below countries topping the list, like the USA with 2.7%, Korea with 2.6% and Canada with 2.5%. Germany was below the 2% goal proclaimed by the EU in its 2006 modernization agenda (Commission of the European Communities 2006), and it was even below the OECD average of 1.5% (OECD 2011: 278).

Fig. 1: Expenditure on educational institutions (tertiary education) as a percentage of GDP in 2008

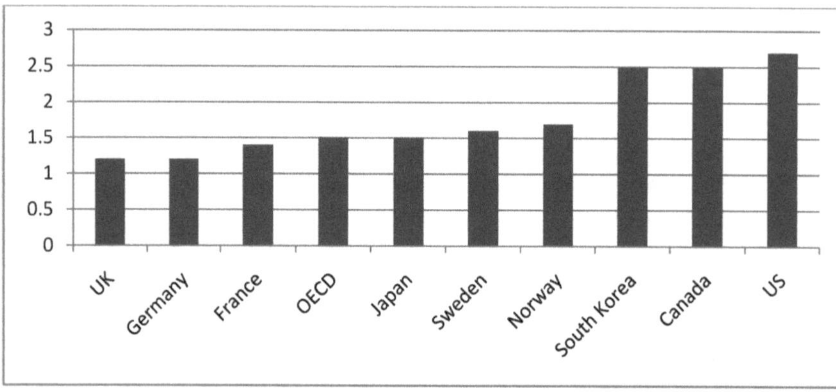

Source: OECD 2011: 229

The percentage of the population with tertiary education differs considerably within the OECD countries, with Germany having a lower participation rate than most comparable OECD countries (in Germany only 26% of the population aged 25 to 34 have received a tertiary degree, while in countries such as Korea, Japan and Canada the respective figure is above 50%, and the OECD average is 37% (OECD 2011: 40). Therefore, total spending per student is a useful additional indicator when comparing countries' investment in higher education (Figure 2). Countries such as the US and Canada top the list. Germany is behind; however, the curve suggests that Germany's spending per student is in line with an increasing overall trend of higher spending on higher education per student with rising per capita income.

Fig. 2: Spending per student in OECD countries

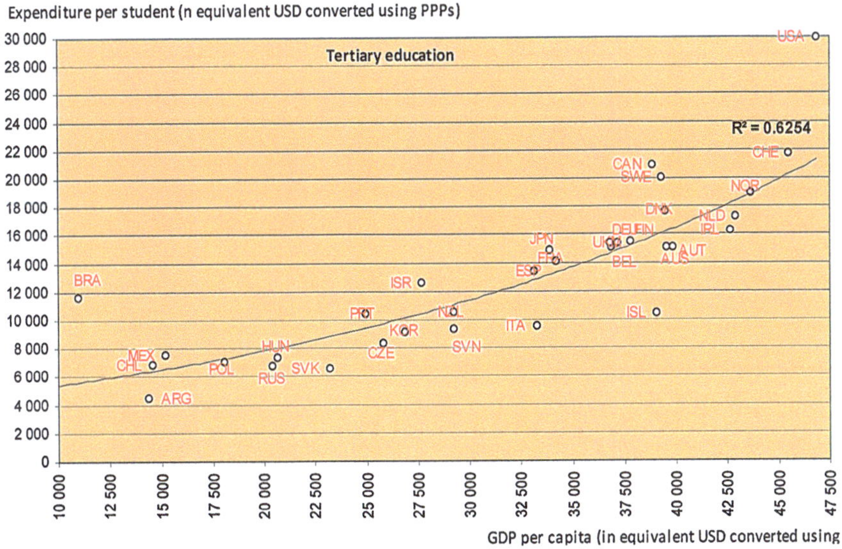

Note: Please refer to the Reader's Guide for the list of country codes used in this chart.
Source: OECD. Argentina : UNESCO Institute for Statistics (World Education Indicators Programme). Tables B1.1a, B1.4 and Annex 2. See Annex 3 for notes *(www.oecd.org/edu/eag2011)*.

Source: OECD 2011: 213

The comparison of funding for tertiary education typically shows considerable differences between those countries where the private sector contributes substantial amounts of money to higher education, and those where it does not.

The three countries with the highest total allocation of funds for tertiary education all have in common the fact that the percentage of private contribution is equal to or above 1% of GDP: South Korea (1.9%), US (1.7%) and Canada (1.0%) (OECD 2011: 231). The OECD average is 0.5% of GDP, and the respective figure for Germany is 0.2%. The following Figure 3 shows the contribution of public and private sources in % of the total contribution. Scandinavian countries' tertiary education systems are characterized by a strong contribution from the public sector and a very small private contribution (however, with a total above-average funding for higher education). Germany is neither characterized by a sizeable private contribution, nor by large public contributions, leading to a total allocation of funds below the OECD average.

Fig. 3: % of public and private spending on higher education in selected OECD countries in 2008

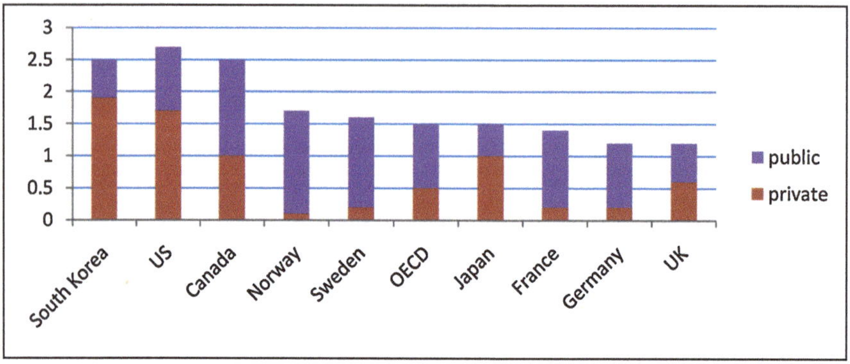

Source: OECD 2011: 231

An analysis of a national system of funding higher education must reflect as well on the specific mix of sources of funding, such as public authorities, international sources, companies/business sector, students, foundations/charities and others, including alumni. The following Figure 4 shows in an abstract way funding sources for higher education institutions.

Fig. 4: Income sources and funding modalities

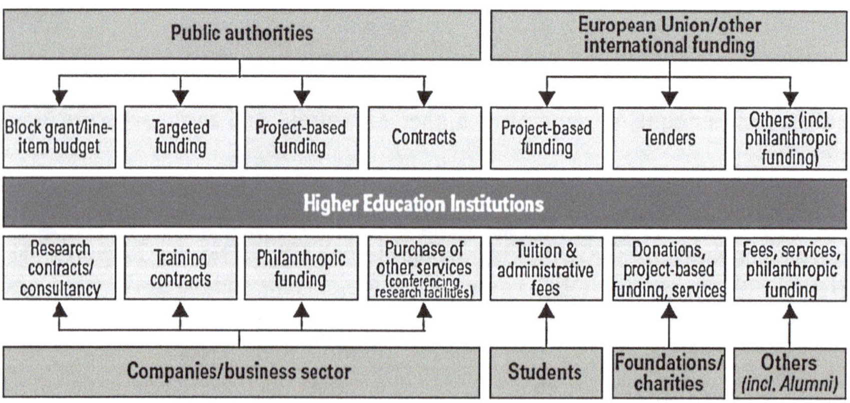

Source: Estermann/Pruvot 2011: 20

Countries differ substantially with regard to the relative importance of these sources of funds. The following Figure 5 uses this overall format to depict income sources for German higher education institutions for 2010 (in million Euros).

Fig. 5: Income sources for German higher education institutions in 2010 in million Euros

<table>
<tr><td colspan="4">Public authorities</td><td colspan="2">European Union</td></tr>
<tr><td colspan="2">Basic funding through states (20 545)</td><td>Third-party funds from states (159)</td><td>Third party funds from federal state (1 308)</td><td colspan="2">Third-party funds: (555)</td></tr>
<tr><td colspan="6">Higher Education Institutions (41 229)</td></tr>
<tr><td>Direct contribution from business (1 246)</td><td>Income from corporate activities (13 471)</td><td>Tuition and administrative fees (1 305)</td><td>German Research Foundation: (2 013)</td><td>Private foundations: Third-party funds: (391)</td><td>Other income: (436) + (235)</td></tr>
<tr><td colspan="2">Companies/business sector</td><td>Students</td><td colspan="2">Foundations</td><td>Others</td></tr>
</table>

Source: Statistisches Bundesamt 2012: 12–17, p. 25

The basic funding through the sixteen states is the most important source of income, contributing 49%. When third-party funds from the states and the federal state are added, the percentage of revenue from national public sources rises to 53%. When the funds coming from the EU and those from the German Research Foundation are added (they almost exclusively come from government), reliance on public sources becomes even more apparent: this figure adds up to 60% of all revenue.

3. A more detailed depiction of funding sources of higher education in Germany

3.1 Public funding of higher education

Federalist tradition in Germany emphasizes the role of the (sixteen) states in the field of higher education: the states are in charge of organizing and funding public tertiary institutions, and of providing for a framework for private higher education institutions. This means that there are sixteen higher education laws in Germany, sixteen ministries for higher education, and sixteen approaches towards funding higher education. And the differences between the states are not marginal. The overall funding for tertiary education differs, the role of the ministries in steering the higher education sector differs, and the number of universities or students per population differs. Some consider this task allocation and diversity a good example of the strength of the educational system in Germany; others see it as problematic, especially considering the small size of some states, as well as the trend towards convergence of systems and approaches, especially in the context of European integration.

Therefore, the role of the federal state in tertiary education is limited. The federal state mainly acts as a coordinator of activities in the sixteen states. And when the federal state provides funding for higher education, the state must seek consensus among and with the states. The German constitution states in Article 91b: "The Federation and the Länder may mutually agree to cooperate in cases of supraregional importance in the promotion of: 1. Research facilities and projects apart from institutions of higher education; 2. Scientific projects and research at institutions of higher education; 3. Construction of facilities at institutions of higher education, including large scientific installations". In such clearly defined areas, an active and financial involvement of the federal state is possible. But this needs to take place in the context of activities of "supraregional importance" and requires approval by the states.

3.1.1 Funding through the federal state

The financial contribution by the German federal government, mainly channeled through the Federal Ministry for Education and Research, aims at supporting research in Germany, developing the physical infrastructure of higher education institutions, providing funds for student loan and scholarship schemes, and providing support for internationalization of higher education in Germany. In terms of overall funding, the contribution stands at 12% of the total public funding for higher education (Statistisches Bundesamt 2012). However, the role of the federal state is not only one of funding activities. Recent initiatives on transforming higher education in Germany have quite often come from the federal government. The Excellence Initiative (an initiative to promote top-level research at German universities), the Higher Education Pact (an initiative to help higher education institutions to deal with additional students), and the Pact for Research and Innovation (a pact for innovation) were highly visible initiatives with a strong role for the federal government.

3.1.2 Funding through the states

In Germany, the (sixteen) states are responsible for covering current expenditure on research and teaching at higher education institutions, and they provide for the general infrastructure (Lohmar/Eckhardt 2011: 83). Public universities receive most of their revenue through such allocations, be they as block grants, or as targeted funding, project-based funding or contracts. 49% of all expenditures on tertiary institutions in Germany in 2010 were covered by the basic allocation through the states ("Grundmittel") (Statistisches Bundesamt 2012: 16 f.). An analysis of indicators such as "total public expenditure on tertiary education institutions", "public spending per student", "public spending as % of GDP", "public spending per inhabitant" shows considerable variations within Germany and the sixteen states.

The public allocation of funds per student in Germany has increased slightly since 1995 (Statistisches Bundesamt 2011: 46). However, behind this overall trend there are significant differences within Germany. Figure 6 below shows the development of the public allocation per student for Germany, for the large states in the

Western part of Germany, and the states in the Eastern part of Germany. It shows the shrinking public allocation per student in the Eastern part of Germany over a period of fourteen years. It also shows that differences in allocation of public funds per student were substantial in 1995, and are much smaller now.

Fig. 6: Public allocation for higher education per student for Germany, Western and Eastern states, 1995–2009

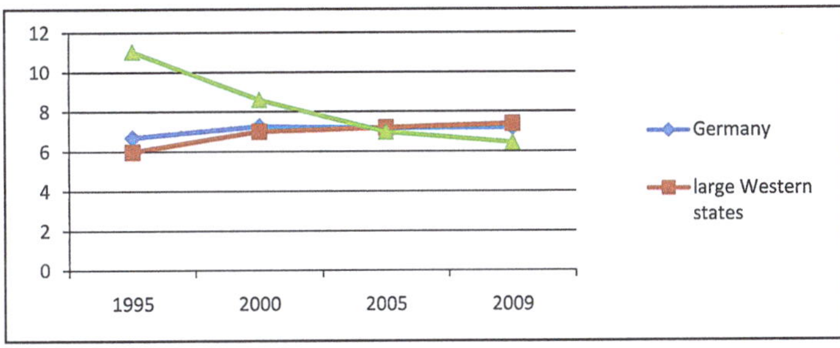

Source: StatistischesBundesamt 2011: 46

However, there are still substantial differences between the sixteen states. The following Figure 7 shows the states ranked according to the same indicator, using figures for 2009. Lower Saxony, with the largest public allocation per student, spends 98% more than Saarland, with the smallest allocation (Statistisches Bundesamt 2011: 44 f.).

Fig. 7: Public spending (basic allocation) per student in 2009

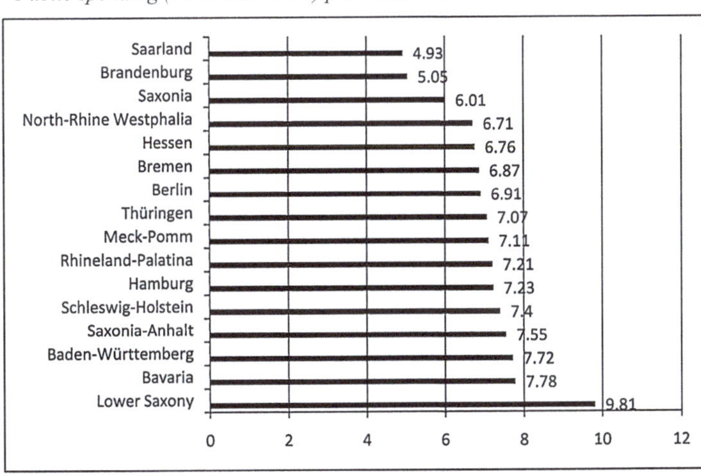

Source: Statistisches Bundesamt 2011: 44 f.

There are a number of factors which can explain the differences, including the disciplinary focus of higher education institutions within a state, the per capita GDP of the states, the debt per capita, the political inclination of the ruling government, or a stronger focus on giving third-party funds and encouraging competition between higher education institutions in the state.

3.2 Third-party funding and the increasing use of competitive mechanisms for the allocation of public funds

The use of third-party funds has gained in significance and importance. In 1995, around 8% of the expenditure of tertiary institutions was covered by third-party funds; in 2010 this figure stood at 14% (Statistisches Bundesamt 2012: 17). Third-party funds are typically allocated through competitive mechanisms, reinforcing a trend which has been in line with a growing appreciation of the value of competitive mechanisms in higher education (Mayer/Ziegele 2009: 54–65).

Fig. 8: Share of funding sources 1995 and 2010

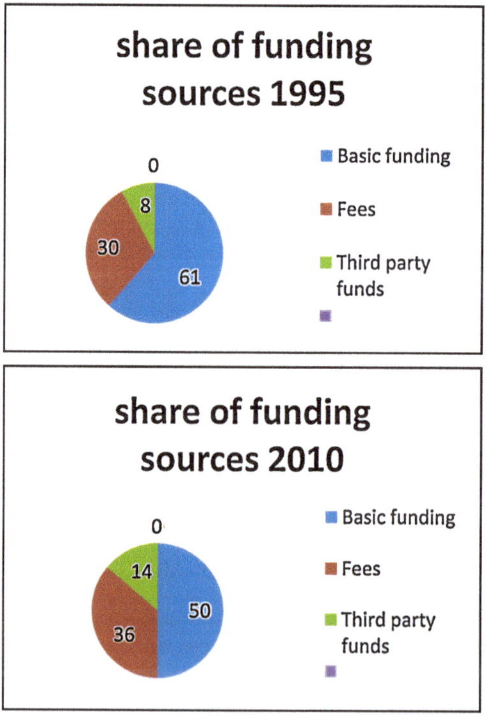

Source: Statistisches Bundesamt 2012: 17

The following graph shows the sources of third-party funds for 2010. The German Research Foundation (DFG) (which mainly receives its funding from the federal ministry) is by far the most important source of funds for higher education institutions: one-third of all funds comes from this key institution for research in Germany.

Fig. 9: Sources of third-party funds in 2010 in % of all third-party funds

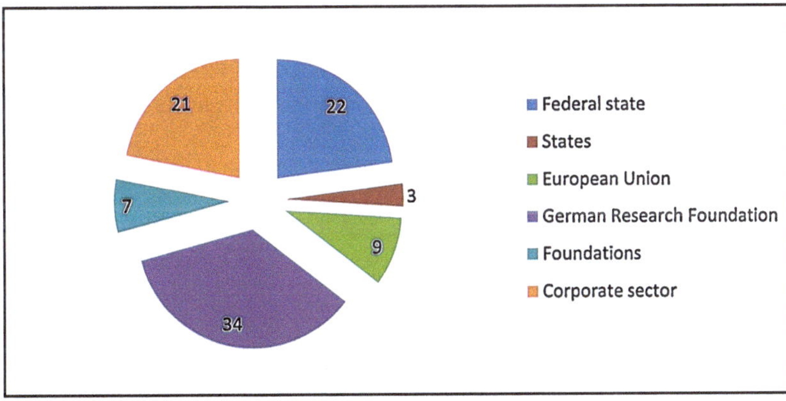

Source: Statistisches Bundesamt 2011: 127

Three disciplines received the lion's share of the third-party funds: mathematics and natural sciences received 26% of all funds, medicine received 24% and engineering 22%. (Statistisches Bundesamt 2012: 25). An analysis of the recipients of third-party funds reveals another bias: higher education institutions in the Southern part of Germany received more third-party funds (in absolute terms and in relative terms, measured for example as third-party funds per professor) than institutions in other parts of Germany (Statistisches Bundesamt 2011: 25).

3.3 Private financing of higher education

Foundations and the corporate sector do provide some funding for higher education institutions. About one-fifth of all third-party funds comes from the corporate sector. And 7% of all third-party funds come from foundations.

The absolute amount of the contribution by the corporate sector has increased over time. However, this kind of funding, very important in many countries, stills plays a minor role in the overall funding structure of German higher education institutions. And the share of the corporate sector's contribution to all third-party funds has declined, from 27% in 2000 to only 21% in 2010 (Statistisches Bundesamt 2011: 121–127).

3.4 Other income for tertiary institutions

Tertiary education institutions rarely run for-profit activities, and do not generate funds from commercial activities. In some cases higher education institutions have opted for an institutional structure where they can generate some funds through professional education activities.

Higher education institutions with hospitals and medical institutions attached, however, do have considerable revenues from their medical activities, which are statistically registered as income from economic activities. Income from such sources was 13.371 billion Euros, which is 33% of all revenues. Of all revenue of this kind, 95% was income by higher education institutions in the field of medicine.

3.5 Administrative and tuition fees

Charging tuition fees for the first academic degree was legally forbidden in Germany until 2005. Students only paid administrative fees of between 100 and 200 Euros per semester. Only students who studied well beyond the regular study period and students of professional programs had to pay tuition fees. A decision by the German Constitutional Court allowed states to introduce tuition fees for regular bachelor and master programs. Seven states subsequently opted for introducing tuition fees. In most cases the tuition fee per semester was 500 Euros.

However, public pressure and claims by critics that tuition fees are unfair and discourage students from low-income families, while denying people a fundamental right, led politicians in five of the seven states to change direction and abolish the tuition fee policy shortly after introducing it. In 2012 only Bavaria and Lower Saxony still require regular students to pay tuition fees.

Figures show that tuition fees were a significant contribution to the overall funding of higher education institutions. Tuition and administrative fees added up to 1,305 million Euros in 2010. Higher education institutions in North Rhine Westphalia received 402 million Euros in 2010 (the state abolished tuition fees only in 2010), the two Southern states Bavaria and Baden-Württemberg both had revenues of 200 million Euros, higher education institutions in Lower Saxony received 135 million Euros (Statistisches Bundesamt 2012: 27). In all these states the financial contribution through tuition and fees was a significant source of income, and, for example, larger than funding through the EU.

3.6 Alumni

Tapping alumni as sponsors of various kinds of activities of tertiary institutions has been uncommon in Germany, and attempts to introduce this kind of funding have failed so far. Many higher education institutions have less than one person working on alumni matters. Fundraising campaigns to approach alumni for donations for their alma mater are not common. Only very few universities can boast of sizeable contributions, typically from a very few donors with large contributions each. The bond between students/graduates and their university is less strong in Germany than in Anglo-Saxon countries, and does not yet involve a readiness to spend considerable amounts of money. This reality feeds back into the way higher education institutions look at investing in staff working in this field. If specialists are hired for contacting alumni and developing and implementing a strategy at all, it is typically not more than one person per institution.

4. The future of funding of higher education in Germany

The following chapter describes two theoretical arguments which can be used when trying to find answers to the two issues concerning the financing of higher education in Germany which are most pressing:

- How much private and how much public funding of tertiary education is necessary and justified?
- What should be the task allocation between the federal state and the sixteen states, including the question of how much funding should be provided by the federal state?

4.1 Higher education and externalities – private versus public financing of higher education

The German higher education system is characterized by a strong bias towards public funding. Key arguments – from a purely economic point of view – to justify public financing of educational activities against private funding by the beneficiaries themselves are the existence of market failures and equity concerns (OECD 2008: 168–171, Kupferschmidt/Wigger 2006: 285–307, Wissenschaftlicher Beirat beim Bundesministerium der Finanzen 2010: 16–24).

Public goods: Public funding of higher education is justified if higher education is a public good. This requires the principles of non-excludability and non-rivalry in consumption to hold. However, neither principle applies for higher education in Germany, as the restricted access to university studies and overcrowding at universities with negative effects on quality clearly show. Therefore, higher education cannot be said to be a public good, and a justification of public funding needs to be based on another line of argumentation.

Existence of externalities: Another reason for public funding of higher education can be the existence of positive externalities: higher education is said to bestow not only benefits on the individual, such as higher income and a lower likelihood of becoming unemployed. Rather, participation in tertiary education is said to bring benefits to society, such as productivity gains and economic growth for the society as a whole, without being fully reflected in higher wages.[1] In addition, it is sometimes argued that higher education has a positive effect on health-consciousness, leading to lower health care costs for society. Furthermore, higher education can have a positive impact on social cohesion, the stability of a society and democracy, driving a wedge between private and social benefit (Autorengruppe Bildungsberichterstattung [eds.] 2012: 199–214).

In Figure 10, a positive externality of higher education is presented by showing that the marginal private benefit (MPB) is below the marginal social benefit (MSB). Assuming constant marginal cost (MC) of higher education there will be a market solution expressed by P* and X*. Efficiency, however, requires equality of MC and MSB. Therefore the quantity of higher education should be increased from X* to X**. This can be achieved by decreasing the price from P* to P**. To this amount, subsidizing of higher education is indicated. Conversely, supposing overcrowding costs occur, it can be shown that the introduction of tuition fees leads to an efficient use of universities (Wellisch 2000: 100–104).

Fig. 10: Positive externality of higher education

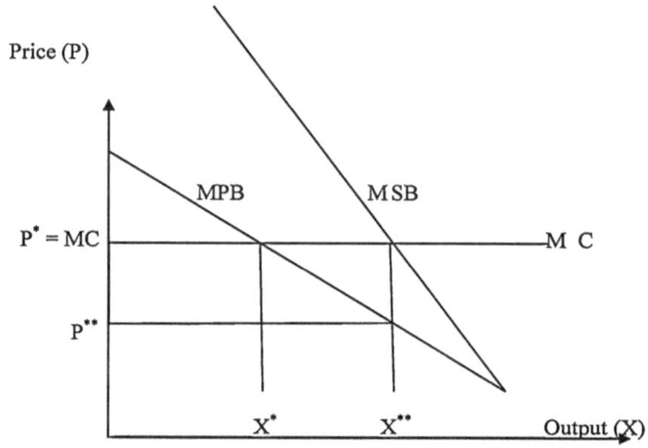

1 The Signaling-Approach (Spence 1973: 355–374) points to the fact that higher education does not contribute to an increase in labor productivity, and rather serves as a signal to employers about individual abilities (Kupferschmidt/Wigger 2006: 293–294).

Empirical evidence for the social benefits of higher education in Germany is difficult to obtain. Graduates of tertiary education institutions tend to retire later and live longer than non-academically trained people. However, this might not be the direct result of a healthier life-style, but rather the result of different kinds of jobs typically held by graduates from universities. And it might be seen mainly as a private benefit. There is no hard evidence of a specific contribution of graduates of tertiary education to democracy and the stability of society. There is some anecdotal evidence that graduates vote far less frequently for radical, especially right-wing parties.

Theory of human capital: Further arguments in favor of public financing of higher education are provided by the theory about human capital. The theory holds that decisions in favor of higher education can be understood as investment decisions, with initial expenses during the period of studies and respective income streams later on, and people making their educational decisions with a calculation of costs and benefits in mind (OECD 2011: 160–175). And if private and social costs and benefits coincide, private educational decisions will maximize not only private welfare, but also social welfare. If social welfare is to be maximized, gifted students with positive rates of return on investments in higher education should have access to loans from the financial system. This is especially important when families do not have savings and other collateral is not available. If financial markets do not provide sound solutions for financing such private and social benefit-enhancing investment, the government might have to step in.

The German system of student loans (BAFÖG) addresses this issue. Students from low-income households have access to a monthly living allowance for the regular course of their studies, half of which needs to be repaid after finishing studies and after graduates have found full employment (Lohmar/Eckhardt 2011: 88–91). Repayment is without interest, and can be stretched over a long period. Still, the burden of debt after graduation is certainly having some negative effect on the decisions taken by potential students.

The issue of having access to bank loans, as discussed in theory and relevant in many countries around the world, is not an issue for students from low-income families. An additional loan scheme has been set up which provides access to loans for students who are not eligible for the abovementioned means-based system of support (Lohmar/Eckhardt 2011: 89).

Information asymmetry concerning future earnings: Underinvestment in tertiary education could also result from the fact that potential students have only limited information about future working conditions, little information about the likelihood of finishing studies successfully, and little information about future earnings (Sachverständigenrat zur Begutachtung der gesamtwirtschaftlichen

Entwicklung 2004, no. 559–562). Risk-averse potential students will therefore discount their prospective income at a rate which is too high, and socially beneficial educational investment decisions will not be taken. If this is the case, the government needs to address this issue.

There has been considerable work and effort in the recent past on calculating the value of higher education. The earning premiums for tertiary education in Germany are substantial. The unemployment rates for graduates are significantly lower than for non-graduates. Therefore, the net present value of tertiary education in Germany is high, the internal rate of return for men and women is substantial (OECD 2011: 131–140). There have been enormous efforts to provide information to the public about career paths after graduating from higher education institutions and about economic benefits. Still, the information might not be available where it is needed most, i.e. in the low income strata of society.

Equity concerns: Equity concerns are important issues for governments, whatever philosophy they finally pursue. Many governments aim at equality of educational opportunity, i.e. making sure that students from low-income families have access to tertiary education without regard to their socio-economic background. The justification for this is only partly economic, but mainly political, social and cultural: societies need mobility across income groups and strata of society. Financing higher education completely through public sources, it is argued, particularly favors students from low-income families because they otherwise shy away from studying, unlike people from middle- or high-income groups: despite loan schemes, scholarships and better information, true upward mobility requires complete coverage of costs for tertiary education by the state. The key argument against this line of thought is based on the assumption that low income groups with a low participation rate in tertiary education indirectly subsidize with their taxes middle- and high-income groups. Empirical studies for Germany have produced mixed results with respect to the long-term success of the current policies (Janeba/Kemnitz/Erhart 2007: 184–205, Borgloh/Kupferschmidt/Wigger 2008: 25–48).

4.2 The theory of fiscal federalism and higher education

A theoretical tool useful in deliberations about the task allocation between the federal and the state level is the theory of fiscal federalism. This theory reflects on the assignment of responsibilities and tasks to the various governmental levels (international, national, state, local). The theory advances a number of arguments which can help decide what level of government is best suited to dealing with specific policy challenges (Oates 1999: 1120–1124, Baldwin/Wyplosz 2012: 99–93 and 473–475).

Different preferences for policies: Preferences for policies are not necessarily homogeneous within a country. The population of some parts of the country might prefer certain policies because of cultural, political, economic or other reasons. If such differences are important, society can maximize welfare by having a decentralized provision of services: a delegation of authority to a lower level might be advisable in order to answer and respond to such differences. A decision on a centralized basis would not satisfy either group. The following graph shows the welfare effects of decentralized and centralized solutions when one group of society, say in the North, has a relatively high willingness to pay (N_2), and another group, say in the South, has a low willingness to pay (N_1). If the central government decides to offer a service according to the Northern preferences (X_2), there is – from a social perspective – an oversupply of services in the South. If the government offers the service according to the preferences of people in the South (X_1), some consumers in the North will be dissatisfied. If the government opts for a compromise and offers a service level (X_C) which is between the maximum and the minimum, there is still a welfare loss.

Fig. 11: Different preferences for higher education

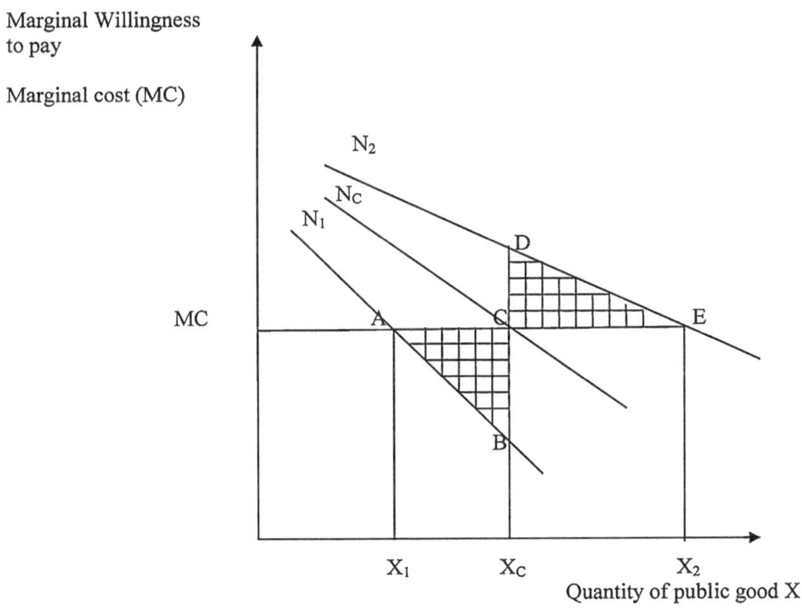

Marginal costs = MC
Willingness to pay of society in state 1 (N_1), state 2 (N_2) and preceived average from the perspective of a centralized government (N_C)

Preferences for providing higher education are heterogeneous. Figure 11 shows that states are characterized by considerable differences in their willingness to allocate funds to higher education institutions. States differ as well in terms of managing universities, i.e. giving them decision-making power over personnel, or using funds discretionarily, or setting priorities. States differ with respect to nurturing private providers of tertiary education. States use their discretion to have disciplinary specializations which respond to their industrial structure, some states invest in disciplines like engineering; others have institutions with a specialization in arts or social sciences. Some states invest in more applied science-oriented institutions, others favor basic research. And states do not implement the Bologna process in the same way. In sum, there is considerable heterogeneity of preferences in higher education in Germany, which is an argument in favor of a more decentralized provision of higher education.

Spillovers and fiscal responsibility: The coincidence of those who pay for the service and those who benefit is an important requirement for decentralized service provisions (Oates 1999: 1120–1123, Olson 1969: 479–483). If spillovers exist and the citizens of a state pay for the provision of a service, but citizens of other states mainly benefit from this service, decentralizing service provision tends to lead to a total provision which is not efficient from the perspective of the society as a whole. Spillovers can be said to exist in higher education if graduates are mobile and work after graduation in a state other than the one where they completed their studies. The argument is based on the fact that state governments also need and use taxes from graduates to provide for higher education spending. Empirical evidence shows that the labor market for graduates of higher education institutions is – with the exception of teachers' education and a few other programs – a national one. Studies found that a sizeable proportion of graduates from tertiary institutions work in states other than the one where they graduated for years after graduation (Huebner 2009: 13). This argument tends to favor a more centralized financing scheme for higher education.

Returns to scale with regard to governmental action: Returns to scale can be relevant for deciding on the most appropriate level of government. If there are positive returns to scale for policy making, a more centralized policy approach might be advisable. Such positive returns to scale can exist when substantial coordination activities are required and a centralized approach allows for reducing such costs. So transaction costs for policy making might matter. On the other hand, if diseconomies of scale exist, coordination becomes increasingly complex and cumbersome and can lead to counterproductive results. In such a case, delegation to the lower level is advantageous. Positive returns to scale are relevant if coordination of higher education policies involves high transaction costs, and if substantial skills are required to keep track of developments, e.g. with respect to the Bologna process or other international developments. The increasing complexity of national and international developments and the shrinking size of ministries in charge of higher education tend to favor centralization. However, past criticism of coordination mechanisms of higher education policies can also be seen as a sign of weakness and a structural deficit of a more centralized provision.

Information asymmetry and task allocation: Information asymmetry is another important argument when deciding upon the task allocation between the federal and state level. Decision makers typically do not have full information about the needs for policy actions, and whatever information is available is not symmetrically distributed. If this is the case, it is reasonable to assume that often the lower level of public authority has more information about conditions, challenges, or alternative paths for actions than the higher level of government, and therefore might be better suited to take decisions. This can justify the delegation of responsibility to the lower level of government.

This argument might be relevant in the field of higher education. Higher education institutions need a supportive local framework, where the local government, the local industry, the local society as a whole supports the development of the academic institutions. Information about such local conditions will be more easily available at the local or state level rather than at the federal level. Such cooperation and therefore prior knowledge about it might be very important in some disciplines. If such information is important, and alternative ways of supplying the federal level with the necessary information and establishing satisfactory decision-making mechanisms do not exist, a case for a more decentralized provision of higher education can be made.

Quality of government: Another important argument is the issue of quality of government and competition between states. Advocates of decentralized solutions argue that communities, provinces, states should compete for good or best solutions. Seen from this perspective, there needs to be a framework which resembles the market mechanism, where a competitive framework for states instills a process of searching for new solutions at the level of government, where creativity is fostered and superiority leads to advantages. This line of reasoning can be applied to higher education: since it is not clear which higher education policy is best for the future of the country, allowing for diversity, allowing for different policies and learning processes might be the advisable approach. In such a system, a competition for best strategies, best policies, best institutional mechanisms might benefit the whole of society.

Principle of subsidiarity: The theoretical considerations mentioned above can and should help decide whether decentralized or centralized solutions are better suited. There is an additional perspective provided by the German philosophical tradition: the principle of subsidiarity. This principle dictates that citizens need to be in charge of their own fate if they are to be able to tackle challenges. A delegation of responsibility to a higher level needs to pass the test of whether the lower level was indeed not able to cope with the challenge. In short: if in doubt, better keep it decentralized. Given the abovementioned ambiguity of arguments applied to higher education, it might be concluded that the principle of subsidiarity suggests leaving higher education in the hands of states.

5. Conclusion

This article has addressed a number of issues. The key findings are summarized in six concluding remarks, containing suggestions for future work:

1. Germany is a country with skill-intensive production. This is where Germany has a comparative advantage and is likely to be successful in the future as well. Skill-intensive production requires an educational system which lays the foundation. A well-funded higher education sector is, therefore, a precondition for future economic success. The gap between total allocation of funds for higher education in some industrialized countries such as the USA or the Scandinavian countries and Germany is significant, and clearly suggests the need for an increase in the total allocation of funds for higher education in Germany. The public debate in Germany does not take sufficient note of this very fundamental fact.
2. An increase in public funding requires a political consensus. The theory of collective action might provide an answer for the lack of political will to increase funding for higher education: students are not well organized, not yet politically active, individual benefits of actively lobbying for higher allocations for education are small. This leads to a situation where students or the young generation are not able to force their interests powerfully into the political process.
3. The comparison between the USA, Japan, Korea on the one hand, and Scandinavian countries on the other hand, shows that countries have a choice concerning the mix of funding sources. Public discussion in Germany should address this issue of designing a funding structure which is compatible with political traditions and sound economic arguments.
4. The discussion of the public and private benefits of higher education has shown that substantial parts of the benefits are private. Furthermore, information gaps and information asymmetries have been reduced: society is now better informed

about the economic benefits of tertiary education than before. Hurdles preventing low income groups from accessing higher education have been reduced. More scholarships are available and a student loan scheme has been established, making it easy for students to get additional finance at subsidized rates. Therefore, from a purely economic perspective, an increase in private funding of higher education seems to be a plausible policy option.

5. If public funds are not increased, and private sources are the only way left to increase total funds, higher education institutions might have to develop mechanisms to attract private funds from foundations, business or alumni, to lobby for funds, to write proposals, to work with donors. This requires new thinking, and a change of culture, both in academia and in the private sector. Private sector donations for teaching and for research, in countries with large contributions from the private sector normally and often interpreted as signs of excellence and attractiveness, do not enjoy such a public response in Germany.
6. The discussion of the key arguments of the theory of fiscal federalism has shown the ambiguity concerning the final responsibility for higher education: the quality of government, the preferences for specific policies and possibly the information asymmetry argument suggest that the states are in charge. The returns to scale and the spillover arguments tend to favor centralization. The principle of subsidiarity might be interpreted in this case as supporting the policies being handled at the state level.

References

Autorengruppe Bildungsberichterstattung (eds.) (2012): *Bildung in Deutschland 2012. Ein indikatorengestützter Bericht mit einer Analyse zur kulturellen Bildung im Lebenslauf*, Bielefeld.

Baldwin, R./Wyplosz, Ch. (2012): *The Economics of European Integration*, 4th ed., London.

Borgloh, S./Kupferschmidt, F./Wigger, B. (2008): Verteilungseffekte der öffentlichen Finanzierung der Hochschulbildung in Deutschland: Eine Längsschnittbetrachtung auf der Basis des Sozioökonomischen Panels. In: *Jahrbücher für Nationalökonomie und Statistik*, Band 228/1, pp. 28–48.

Commission of the European Communities (2006): Delivering on the modernization agenda for universities: education, research and innovation, 2006, online: http://eur-lex.europa.eu/LexUriServ/LexUriServ.do?uri=COM:2006:0208: FIN: EN:PDF, retrieved on 2nd of September 2012.

Estermann, T./Pruvot, E. (2011): European Universities Diversifying Income Streams: an overview of the study. In: *Bayerisches Staatsinstitut für Hochschulforschung und Hochschulplanung: Beiträge zur HOCHSCHULFORSCHUNG* 2/2011, pp. 18–38.

Huebner, M. (2009): *Essays in Fiscal Federalism – Welfare competition and the funding of higher education*, Mannheim.

Janeba, E./Kemnitz, A./Ehrhart, N. (2007): Studiengebühren in Deutschland: Drei Thesen und ihr empirischer Gehalt. In: *Perspektiven der Wirtschaftspolitik*, Vol. 8, Issue 2, pp. 184–205.

Kupferschmidt, F., Wigger, B. U. (2006): Öffentliche versus private Finanzierung der Hochschulbildung: Effizienz- und Verteilungsaspekte. In: *Perspektiven der Wirtschaftspolitik*, Bd. 7/2, pp. 285 – 307.

Lohmar, B./Eckhardt, T. (2011): *The Education System in the Federal Republic of Germany 2010/11*, Bonn.

Mayer, P./Ziegele, F. (2009): Competition, Autonomy and New Thinking: Transformation of Higher *Education in Federal Germany. In:* Higher Education Management and Policy, Vol. 21/2, pp. 1–20.

Oates, W. E. (1999): An Essay on Fiscal Federalism. In: *Journal of Economic Literature*, Vol. 37, No. 3, pp. 1120–1149.

OECD (2008): *Tertiary Education for the Knowledge Society*. Volume 1 and Volume 2, OECD Publishing, Paris.

OECD (2011): *Education at a glance 2011: OECD indicators*, OECD Publishing, Paris.

Olson, M. (1969): The Principle of "Fiscal Equivalence". The Division of Fiscal Responsibilities between Different Levels of Government. In: *American Economic Review*, 59, pp. 479 - 487.

Sachverständigenrat zur Begutachtung der gesamtwirtschaftlichen Entwicklung (2004): *Erfolge im Ausland – Herausforderungen im Inland*, Jahresgutachten 2004/05, Wiesbaden.

Spence, A. M. (1973): Job Market Signalling. In: *Quarterly Journal of Economics*, Vol. 87, pp. 355–374.

Statistisches Bundesamt (2011): *Bildung und Kultur – Monetäre hochschulstatistische Kennzahlen 2009* – Fachserie 11, Reihe 4.3.2., Wiesbaden.

Statistisches Bundesamt (2012): *Bildung und Kultur – Finanzen der Hochschulen – Fachserie 11*, Reihe 4.5. Wiesbaden.

Watts, R. and P. Hobson (2000): Fiscal Federalism in Germany, retrieved 3rd of September2012: http://www.aucc.ca/_pdf/english/programs/cepra/watts_hobson.pdf

Wellisch, D. (2010): *Finanzwissenschaft I. Rechtfertigung der Staatstätigkeit*, München.

Wissenschaftlicher Beirat beim Bundesministerium der Finanzen (2010): *Beitragspflichtiger Hochschulzugang*, Berlin.

Contributors

Hans Adam is Professor of Economics at the University of Applied Sciences Osnabrück. He studied economics at the Technical University of Berlin and received his doctorate from the University of Hannover. His major field of research is public finance.

Robyn Hannigan is currently the Chair of the Environmental, Earth and Ocean Sciences Department at UMass Boston. Prior to this she was a program officer at the National Science Foundation in the Division of Biological Infrastructure. She currently serves as the Chair of Consortium of Universities for the Advancement of Hydrologic Sciences, Inc., Member-at-Large on the Executive Committee of the National Council for Science and the Environment-Council of Environmental Deans and Directors, and Chair of the National Science Foundation Biological Sciences Research for Undergraduate Students Leadership Council. Dr. Hannigan is an internationally reputed leader in geochemical research with over 100 publications.

Eunsook Hyun, Ph.D., is Professor of Curriculum Studies and Associate Provost at the University of Massachusetts, Boston, USA. Dr. Hyun is an established scholar and experienced administrator in higher education. She has widely published in many areas, including curriculum theorizing, transdisciplinary higher education curriculum, developmentally and culturally appropriate practice (DCAP), early childhood education, teacher education, critical pedagogy, gender studies, bilingual education, theory of teacher reflectivity and multiple/multiethnic perspective-taking. Now Dean and Professor of Curriculum Studies at the Charter College of Education, California State University, Los Angeles.

Winston E. Langley is Provost and Vice Chancellor for Academic Affairs, University of Massachusetts Boston. He is a professor of Political Science and International Relations, with additional degrees in biology, diplomatic history and law. His scholarly interests include human rights, especially as they pertain to women and children, but also the legal and moral order on which those rights are grounded, and the implications of that order for our collective future. Nuclear weapons proliferation, environmental degradation, the global political economy, and the evolution of intergovernmental organizations, especially the United Nations, and the European Union, and ASEAN, are also part of his scholarly and teaching emphases. He has published widely, in his field, with six books and over 70 articles, among other published works.

Peter Mayer studied business administration and economics at the Universities of Frankfurt and Milwaukee, and completed his doctorate degree at the Johann Wolfgang Goethe University of Frankfurt. After working for Friedrich-Ebert-Foundation in Bonn, Accra /Ghana and Seoul/ South Korea, he became Professor for International Economics at the University of Applied Sciences Osnabrück. One of his research interests is higher education management.

Rainer Paslack is a sociologist, philosopher and graduate human biologist. He is currently a research assistant at the Medical School of Hannover (Germany). He is the author of several books and over 40 articles in the life sciences, especially biotechnology, bioethics and environmental ethics. He is also co-editor of the series *Medicine, Technology and Society*.

Robert F. Rich recently retired from the position of Director of the University of Illinois Institute of Government and Public Affairs and Professor of Law, Political Science, Medical Humanities and Social Sciences, Community Health, and Health Policy and Administration. He is the author of eight books and over 50 articles in the areas of health law and policy, federalism, information policy, and science and technology policy. He is also the co-editor of the new *Encyclopedia on Health Services Research* published in May of 2009.

Jürgen Robienski studied law in Germany and Sweden. Since 1996 he has been working as a lawyer with offices in Hannover and Müden (Germany), and since 2010 he has also been a research associate at Hannover Medical School. His research concerns medical and economic law in general, especially legal questions of bio-banking, human tissue, gene diagnosis, reproductive and personalized medicine, incidental findings, gene doping, enhancement and synthetic biology. He has published several articles and books on these topics.

Jürgen W. Simon studied Law and Economics in Tübingen, Freiburg und Munich; he gained a Dr. jur. from the University of Freiburg and a Habilitation degree in Civil Law, Economic Law and Legal Computer Science from the University of Hanover. He then researched and taught Medical and Environmental Law. For many years he was Director of the Institute of Legal Sciences at the Leuphana University Lüneburg; he is currently a Visiting Scholar at the University of London, a member of the Center for Ethics and Law of the Life

Sciences (CELLS) at Hanover Medical School and a member of the IALS Working Group "Public Policy & Life Sciences".

Dan Simovici obtained his PhD in mathematics from the University of Bucharest, Romania after he got an MS in Electrical Engineering. Currently, he is a professor of Computer Science and Graduate Program Director at the University of Massachusetts Boston. He is the author or co-author of eighteen books and 150 research papers. His main research interests are in data mining and its applications in biology and medicine; he is also interested in multi-valued logic and serves as editor-in-chief for the Journal of Multiple-Valued Logic and Soft Computing.

Heidrun Sturm studied medicine in Bochum, Hannover and Baltimore, she completed a Masters in Public Health at Hannover Medical School and a PhD on comparative healthcare research at the University of Groningen, NL. She has worked, amongst other things, as a researcher at Yale University (USA) and as general medical manager of the Comprehensive Cancer Center in Tübingen. Today she is a healthcare consultant working for the pharmaceuticals industry, health insurance companies and other players in the healthcare arena. She is also a member of the board of the German Patient Association (DGVP).

Julian Ziegler studied Law at the Free University Berlin and, following an academic year program at the Institute of Government and Public Affairs at the University of Illinois at Urbana-Champaign, took a Dr. iur. degree at the Law Faculty of Humboldt Universität zu Berlin. He is currently working in the real estate department of a German retail company.

MEDIZIN, TECHNIK UND GESELLSCHAFT
MEDICINE, TECHNOLOGY, AND SOCIETY

Herausgegeben von Rainer Paslack, Frank Böker und Hilmar Stolte
Edited by Rainer Paslack, Frank Böker and Hilmar Stolte

Bd./Vol. 1 Kurt Bayertz / Hilmar Stolte / Ursula Zimmermann (eds.): Ethical dimensions of Technology Transfer in Biomedicine. 1993.

Bd./Vol. 2 Rainer Paslack / Hilmar Stolte (Hrsg.): Gene, Klone und Organe. Neue Perspektiven der Biomedizin. 1999.

Bd./Vol. 3 Julian Ziegler: Genanalysen in der Deutschen Privaten Krankenversicherung. 2008.

Bd./Vol. 4 Rainer Paslack: Somatische Gentherapie – Eine historische Fallstudie (1965-2000). Von der Forschung zur klinischen Anwendung: Geschichte, Regulation und Bewertung. 2009.

Bd./Vol. 5 Eunsook Hyun / Rainer Paslack / Hilmar Stolte (eds.): Transdisciplinary Interfaces in the Life Sciences. 2014.

www.peterlang.com

FSC
www.fsc.org
MIX
Papier | Fördert
gute Waldnutzung
FSC® C083411

Kolibri 360 GmbH
Ferdinand-Jühlke-Straße 7
99095 Erfurt, Deutschland
produktsicherheit@kolibri360.de